TC 3-19.5
November 2009

Nonlethal Weapons Training

DISTRIBUTION RESTRICTION: Distribution authorized to DOD and DOD contractors only to protect technical or operational information from automatic dissemination under the International Exchange Program or by other means. This determination was made on 6 November 2006. Other requests for this document must be referred to the Commandant, USAMPS, ATTN: ATSJ-L, 401 MANSCEN Loop, Suite 1088, Fort Leonard Wood, Missouri 65473-8929.

DESTRUCTION NOTICE: Destroy by any method that will prevent disclosure of contents or reconstruction of the document.

Headquarters, Department of the Army

This publication is available at
Army Knowledge Online (www.us.army.mil) and
General Dennis J. Reimer Training and Doctrine
Digital Library at (www.train.army.mil).

Training Circular
No. 3-19.5

TC 3-19.5
Headquarters
Department of the Army
Washington, DC, 5 November 2009

Nonlethal Weapons Training

Contents

Page

Preface vii

Chapter 1	**Introduction**	1-1
	Purpose	1-1
	Scope	1-1
	Background	1-1
	Policy for Nonlethal Weapons	1-2
	Nonlethal Capability Requirements	1-2
	Nonlethal Weapons Support in Army Operations	1-2
	Nonlethal Multi-Media Training Support Package	1-3
	Joint Nonlethal Weapons Educational Resources	1-3
Chapter 2	**Conduct of Training**	2-1
	Training Strategy	2-1
	User Training	2-1
	Nonlethal Munitions Training Requirements	2-1
	Certification and Familiarization Training	2-2
	Composite Risk Assessment	2-4
	Recommended Range Procedures	2-5
	Rules of Engagement	2-6
	Doctrinal Guidance for Nonlethal Weapons	2-8
Chapter 3	**Employ 12-Gauge Point and Area Target Munitions**	3-1
	Overview	3-1
	M1012 12-Gauge Nonlethal Point Round Specifications	3-1

DISTRIBUTION RESTRICTION: Distribution authorized to DOD and DOD contractors only to protect technical or operational information from automatic dissemination under the International Exchange Program or by other means. This determination was made on 6 November 2006. Other requests for this document must be referred to the Commandant, USAMPS, ATTN: ATSJ-L, 401 MANSCEN Loop, Suite 1088, Fort Leonard Wood, Missouri 65473-8929.

DESTRUCTION NOTICE: Destroy by any method that will prevent disclosure of contents or reconstruction of the document.

Contents

        M1012 Employment Considerations 3-2
        M1012 Certification Standards 3-2
        M1012 Sighting 3-3
        M1013 12-Gauge Crowd Dispersal Cartridge Specifications 3-3
        M1013 Employment Considerations 3-4
        M1013 Familiarization Standards 3-4
        M1013 Sighting 3-5

Chapter 4   **Employ 40 Millimeter Nonlethal Munitions 4-1**
        Overview 4-1
        M1006 Sponge Grenade Specifications 4-1
        M1006 Employment Considerations 4-2
        M1006 Certification Standards 4-2
        M1006 Sighting 4-2
        M1029 40MM Nonlethal Crowd Dispersal Cartridge Specifications 4-3
        M1029 Employment Considerations 4-4
        M1029 Familiarization Standards 4-5
        M1029 Sighting 4-5

Chapter 5   **Employ Nonlethal Grenades 5-1**
        Overview 5-1
        M84 Stun Hand Grenade Specifications 5-1
        M84 Employment Considerations 5-2
        M84 Safety Considerations 5-2
        Arm M84 Stun Hand Grenade 5-3
        M104 Sting Ball Grenade Specifications 5-4
        M104 Employment Considerations 5-5
        M104 Safety Considerations 5-5
        Arm M104 Sting Ball Grenade 5-5

Chapter 6   **Employ Modular Crowd Control Munition 6-1**
        Overview 6-1
        M5 MCCM Specifications 6-1
        M5 MCCM Employment Considerations 6-4
        M5 MCCM Safety Considerations 6-4
        Emplacing, Aiming, and Arming the M5 MCCM 6-5
        Misfire Procedures 6-6
        Handling and Recovery Procedures 6-7
        Practical Exercise 6-8

Contents

Chapter 7	Employ 66 Millimeter Nonlethal Grenades	7-1
	Overview	7-1
	66MM Nonlethal Grenade Specifications	7-1
	Nonlethal Grenade Loading and Unloading Procedures	7-2
	Nonlethal Grenade Launching Procedures	7-3
Chapter 8	Employ Individual Riot Control Agent Disperser	8-1
	Overview	8-1
	Definitions	8-1
	M39 IRCAD Specifications	8-2
	IRCAD Employment Techniques	8-4
	IRCAD Safety Considerations	8-6
	Decontamination, Recovery, and First Aid	8-6
	Characteristics of OC	8-7
	Physical and Psychological Effects of OC	8-8
	Environmental and Storage Considerations	8-9
Chapter 9	Employ Electro-Muscular Disruption Device	9-1
	Overview	9-1
	Electro-Muscular Disruption Specifications	9-1
	Loading and Unloading Procedures	9-9
	Preventive Maintenance, Checks, and Services	9-10
	X26E Firing Procedures	9-13
	Probe Removal	9-15
	Download Usage Data	9-15
Chapter 10	Employ Vehicle Barriers and Arresting Devices	10-1
	Overview	10-1
	Portable Vehicle Arresting Barrier Specifications	10-1
	PVAB Site Selection	10-1
	Bump Module Assembly	10-2
	Brake Box Installation	10-3
	Anchor Installation	10-4
	Erector Assembly	10-6
	Net Assembly	10-6
	Vehicle Light Arresting Device Specifications	10-10
	VLAD Site Selection	10-11
	VLAD Rapid Deployment	10-11
	VALD Lanyard Deployment	10-12
	Remove Net From Vehicle	10-12
	Rapid Deploy Vehicle Denial Systems	10-13

Contents

Chapter 11	Employ Communication Devices	11-1
	Overview	11-1
	Phraselator Specifications	11-1
	Phraselator Employment Considerations	11-2
	Operate Phraselator	11-2
	Phraselator Operator Maintenance	11-7
	Demonstrate Phraselator Operation	11-8
	Voice Response Translator Specifications	11-9
	VRT Employment Considerations	11-9
	Operate VRT	11-9
	Demonstrate VRT Operation	11-11
	Magnetic Audio Device Specifications	11-11
	MAD Employment Considerations	11-12
	Operate MAD	11-13
	Demonstrate MAD Operation	11-15
Appendix A	Nonlethal Capabilities Set	A-1
Appendix B	Nonlethal Munitions Practical Exercise	B-1
Appendix C	M5 MCCM Practical Exercise	C-1
Appendix D	X26E TASER Performance Evaluation	D-1
Appendix E	Components of the Checkpoint Operations Mission Module	E-1
Glossary		Glossary-1
References		References-1
Index		Index-1

Figures

Figure 2-1	Nonlethal Munitions Shot Placement Zones	2-2
Figure 2-2	Force Continuum	2-7
Figure 2-3	Doctrinal and Training Resources	2-8
Figure 3-1	Cut-Away View of the M1012 12-Gauge Point Cartridge	3-1
Figure 3-2	M1012 Sighting at 10-20 Meter Targets	3-3
Figure 3-3	Cut-Away View of the M1013 12-Gauge Crowd Dispersal Cartridge	3-4
Figure 3-4	M1013 Sighting at 10-20 Meters	3-5
Figure 4-1	M1006 Sponge Grenade Cut-Away View	4-1
Figure 4-2	M1006 Sighting at 50 Meters	4-3
Figure 4-3	M1006 Sighting at 10 to 45 Meters	4-3
Figure 4-4	M1029 Cut-Away View of the M1029	4-4
Figure 4-5	M1029 Sighting at 15 to 30 Meters	4-5
Figure 4-6	M1029 Sighting at 10 to 15 Meters	4-5
Figure 5-1	M84 Stun Grenade	5-1
Figure 5-2	M84 Stun Grenade Right Handed Procedure	5-3
Figure 5-3	M84 Stun Grenade Left Handed Procedure	5-4
Figure 5-4	Cut-Away View of the M104 Sting Ball Grenade	5-4
Figure 6-1	M5 MCCM with Components	6-1
Figure 6-2	M81 Igniter	6-3
Figure 8-1	M39 Individual Riot Control Agent Disperser	8-3
Figure 9-1	X26E Parts and Functions	9-1
Figure 9-2	X26E with Holster	9-2
Figure 9-3	Cut-Away View of the X26E Cartridge	9-6
Figure 9-4	X26E Probe Types	9-8
Figure 9-5	Results of Improper Handling Technique	9-10
Figure 10-1	Site Requirements for PVAB	10-2
Figure 10-2	Bump Module Assembly	10-3
Figure 10-3	Brake Box Installation	10-4
Figure 10-4	Anchor Installation	10-5
Figure 10-5	Net Assembly	10-7
Figure 10-6	Pack Net	10-9
Figure 10-7	Pack Net Continued	10-9
Figure 10-8	VLAD System Packed for Transport	10-10
Figure 10-9	VLAD Positioned Across Roadway	10-11
Figure 10-10	Magnum Spike System	10-13
Figure 10-11	Caltrops	10-13
Figure 11-1	Phraselator Features and Controls	11-1

Contents

Figure 11-2 Features of the Magnetic Audio Device 11-12
Figure 11-3 MAD Set Up Procedures 11-14

Tables

Table 2-1   Effects of Nonlethal Munitions at Distances 2-3
Table 2-2   40MM Sponge Grenade and 12-GA Point Round Individual Training Program 2-4
Table 2-3   40MM Crowd Dispersal and 12-GA Crowd Dispersal Individual Training Program 2-4
Table 3-1   Collective Tasks for Annual Nonlethal Weapons Training 3-2
Table 10-1 Vehicle Stopping Distances 10-11
Table 11-1 Approximate Ratios of Recording Time to Storage Used 11-6
Table B-1   Nonlethal Munitions Qualification Range Checklist B-5
Table B-2   Nonlethal Qualification Score Card B-6
Table D-1   TASER Qualification Range Checklist D-2

Preface

This Training Circular (TC) serves as a training support package for nonlethal training. The TC provides guidance for leaders conducting nonlethal weapons (NLW) training. It focuses on effective use of NLW and capabilities.

This TC applies to the Active Army, the Army National Guard (ARNG)/National Guard of the United States (ARNGUS), and the United States Army Reserves (USAR) unless otherwise stated.

TC 3-19.5 is organized into the following eleven chapters and five appendices that provide nonlethal weapons training strategies and training support materials:

- Chapter 1, Introduction, explains the purpose and scope of the TC and describes the Army's nonlethal capability requirements.
- Chapter 2, Conduct of Training, provides the training strategy for nonlethal weapons training.
- Chapter 3, Employ 12 Gauge Point and Area Target Munitions, describes the characteristic, employment techniques, and training for the 12 gauge nonlethal rounds.
- Chapter 4, Employ 40 Millimeter Nonlethal Munitions, describes the characteristic, employment techniques, and training for the 40mm nonlethal munitions.
- Chapter 5, Employ Nonlethal Grenades, describes the specifications, safety features, and employment considerations for nonlethal hand thrown grenades.
- Chapter 6, Employ Modular Crowd Control Munition, explains how to train Soldiers to emplace, aim, and arm the M5 Modular Crowd Control Munition.
- Chapter 7, Employ 66 Millimeter Nonlethal Grenades, describes the loading, unloading, and launching procedures for 66mm nonlethal grenades.
- Chapter 8, Employ Individual Riot Control Agent Disperser, provides employment procedures for the M39 individual riot control agent disperser and describes physical and psychological effects of riot control agents.
- Chapter 9, Employ Electro-Muscular Disruption Device, explains the function, characteristics, and employment of the X26E TASER®.
- Chapter 10, Employ Vehicle Barriers and Arresting Devices, describes training and employment of portable nonlethal vehicle barriers, arresting devices, spike systems, and caltrops.
- Chapter 11, Employ Communication Devices, explains the techniques for operating handheld language translators and voice amplifier.
- Appendix A, Nonlethal Capabilities Set, list the components currently available in the nonlethal capabilities set.
- Appendix B, Nonlethal Munitions Practical Exercise, contains instructor guidance and firing commands for conducting a practical exercise on 12 gauge and 40mm nonlethal munitions.
- Appendix C, M5 MCCM Practical Exercise, provides instructor guidance and firing commands for conducting a practical exercise on the M5 Modular Crowd Control Munition.

Preface

- Appendix D, X26E TASER® Performance Evaluation, provides a training and qualification standard for TASER training.
- Appendix E, Components of the Checkpoint Operations Mission Module, describes the various nonlethal components and devices available to support checkpoint, roadblock, and access control operations.

The proponent for this publication is the United States Army Training and Doctrine Command (TRADOC). Send comments and recommendations on Department of the Army (DA) Form 2028 (*Recommended Changes to Publications and Blank Forms*) directly to Commandant, United States Army Military Police School. ATTN: Army Nonlethal Scalable Effects Center, 401 MANSCEN Loop, Suite 1088, Fort Leonard Wood, Missouri 65473. Submit electronic DA Form 2028 or comments and recommendations on DA Form 2028 format by email to LEON.USAMPSNCE@conus.army.mil

Chapter 1

Introduction

PURPOSE

1-1. TC 3-19.5 provides guidance on specific NLW training with emphasis on User Training, Train-the-Trainer Training, and Unit Training. It is designed to be used with FM 3-22.40, *Multi-Service TTP for the Tactical Employment of Nonlethal Weapons*, and the Multi-Media Training Support Package (MMTSP). The MMTSP is a Warrior TSP designed to train individual tasks.

SCOPE

1-2. TC 3-19.5 is designed to train leaders and Soldiers, primarily at company-level and below, on the tactical employment of NLW in support of operating forces.

1-3. Leader training for Field Grade Officers, though not included in this publication, consists of formulating nonlethal policies, establishing the appropriate rules of engagement (ROE), and deciding when the use of nonlethal capabilities is best suited.

BACKGROUND

1-4. United States Forces will respond to a myriad of situations across the range of military operations. At the same time, the military will face increased media attention, worldwide environmental concerns, and a low national tolerance for long, lethal, and costly campaigns even where vital interests of the nation are clearly defined. Nonlethal capabilities can expand options and tools available to Commanders.

1-5. NLW enhance the Army's ability to meet requirements of applying force proportional to the threat and discriminating in the application of force during military operations. Nonlethal capabilities can reduce the risks of perceived excessive force, promote international political support, alleviate environmental concerns, enhance post conflict transitions and termination, and reduce the cost of post conflict reconstruction.

1-6. Many situations that begin as standoffs have the potential to escalate to a point where lethal force may be justified. However, early and aggressive use of NLW may prevent many of these types of situations from escalating to deadly force levels. In this aspect NLW can reduce Soldier and non-combatant injuries by stopping threats from a safe distance.

1-7. It must be made clear that, NLW are not meant to be a substitute for lethal force. Commanders must ensure that NLW are not deployed without lethal support in the form of lethal overwatch. Employing lethal overwatch in support of NLW benefits the commander in several ways. First, lethal overwatch provides Soldiers employing NLW with confidence in knowing that if the situation were to escalate lethal fire is immediately available. Additionally, lethal overwatch becomes the on-scene eyes and ears of the commander providing valuable information that could influence future actions and planning.

1-8. NLW and capabilities provide a valuable set of tools for a variety of missions, particularly when deadly force is not authorized or the preferred course of action. If used effectively, NLW can contribute to mission accomplishment and:

- Limit destruction.
- Reduce fratricide.
- Limit civilian casualties.
- Reduce the likelihood for conflict escalation.
- Enhance protection

Chapter 1

- Reduce reconstruction cost.
- Gain public trust and acceptance.
- Provide greater range of graduated response options.

POLICY FOR NONLETHAL WEAPONS

1-9. The Army proponent for NLW policy is The Office of the Provost Marshal General. All question regarding NLW policy, both tactical and non-tactical should be directed to The Office of the Provost Marshal General, Military Police Policy Division, 2800 Army Pentagon, Washington, DC 20310-2800.

NONLETHAL CAPABILITY REQUIREMENTS

1-10. The tools that make up nonlethal capabilities are broken down into two groups: Counter-Personnel and Counter-Materiel. Capabilities that are considered Counter-Personnel characteristically influence behavior and activities of a potentially hostile crowd, incapacitate personnel, aid in the seizure of personnel and deny personnel access to key areas. Capabilities considered Counter-Materiel provide the ability to disable or neutralize vehicles or facilities without destroying them. Counter-Materiel capabilities can also deny vehicle access to certain areas or facilities.

1-11. Counter-Personnel Capabilities. Counter-personnel capabilities currently consist of rounds such as 12-gauge and 40 millimeter point and area rounds, chemical irritants, impact weapons, and electro muscular disruption devices such as TASER®.

1-12. Counter-Materiel capabilities currently consist of vehicle stopping / arresting devices. Some of the items available are the Portable, Vehicle Arresting Device (M1, PVAB), the Vehicle Lightweight Arresting Device (M2, VLAD) and an array or caltrops and spike strips. The above listed and other Counter-Personnel and Counter-Materiel devices can be found in the current Nonlethal Capabilities Set (NLCS). Appendix A provides a detailed listing of NLCS items.

1-13. The NLCS is designed to equip a Brigade size element with all the appropriate equipment to carry out its basic missions in accordance with current Rules of Engagement (ROE) and Escalation of Force (EOF) procedures. The NLCS is an updated version of the Platoon NLCS which contained mainly riot control gear. These kits are more mission specific and are tailored to a specific purpose.

1-14. A NLCS contains the equipment required to satisfy most operational requirements for an enhanced capability to apply nonlethal force. It is designed to augment lethal forces and will be employed in a manner that will incapacitate personnel or material, while minimizing fatalities or permanent injury or damage to property and the environment.

NONLETHAL WEAPONS SUPPORT IN ARMY OPERATIONS

1-15. NLW support all military operations from peacetime military engagement to major combat operations. Nonlethal capabilities have particular application in civil support operations where there is the need to enhance the Army's ability to meet requirements of applying force proportional to the threat and the desire to protect non-combatants, promote international political support, alleviate environmental concerns, and enhance post conflict transitions and termination.

1-16. Specific missions where NLW complement lethal force include the following:

- Convoy security in urban operations where population density and characteristics of the area require the careful employment of force to minimize loss of life and destruction of property.
- Crowd control and civil disturbance operations where nonlethal munitions, protective equipment, and support equipment enable Army elements to execute population control missions and provide protection while minimizing collateral damage and injury to personnel.
- Cordon and search operations where friendly forces come in close contact with the local population under stressful and often hostile circumstances.

- Checkpoint operations where nonlethal capabilities protect friendly forces and improve public opinion and acceptance.
- In Detainee Operations where all displays of violence must be brought under control quickly while avoiding unnecessary injury or death to the detainees.

1-17. When Commanders integrate NLW training into unit training, field training exercises, and situational training Soldiers gain the confidence and skills necessary for the proper employment of NLW.

NONLETHAL MULTI-MEDIA TRAINING SUPPORT PACKAGE

1-18. The Army Nonlethal Scalable Effects Center (ANSEC) at Fort Leonard Wood, Missouri, develops and implements overarching nonlethal strategies that address solutions to identified gaps in required Army mission tasks. The nonlethal MMTSP is one of the critical tools used to ensure both the success of today's Soldier and that of the future operating force.

1-19. The MMTSP focuses on the Soldier's use and deployment of the NLCS. Development of the MMTSP included the assessment and selection of required tasks, development of lesson plans, training materials and aids, student courseware, and student assessment tools. The MMTSP target audience is at the squad and team leader level.

1-20. To ensure Soldiers are properly trained, equipped, and educated in the application of nonlethal capabilities, ANSEC deploys Mobile Training Teams. Mobile Training Teams are currently being deployed world-wide to assist units with their nonlethal training needs. The MMTSP serves as the primary training tool used by the Mobile Training Teams and is included in the NLCS.

1-21. To ensure qualified nonlethal instructors are properly managed, ANSEC requested and U.S. Army Human Resources Command established an Additional Skill Identifier (ASI) 2A (Nonlethal Advisor) to identify all Army Soldiers, enlisted or officer, who are Inter-Service Nonlethal Individual Weapons Instructor Course (INIWIC) trained and qualified to be NLW instructors. Soldiers possessing an ASI 2A should be used to assist planners at Brigade and Division levels on the employment of NLW and the integration of nonlethal capabilities into training.

JOINT NONLETHAL WEAPONS EDUCATIONAL RESOURCES

1-22. The Joint Nonlethal Weapons Program uses a wide range of educational programs to promote increased understanding of non-lethal weapons and technologies. A web-based online course taught through Penn State University Fayette allows interested military and government personnel to increase their knowledge of nonlethal technology through the internet (http://www.fe.psu.edu/CE/23514.htm). The latest version of the course is available on a single DVD for those personnel who may have limited internet access.

1-23. The Penn State's Nonlethal Weapons: Policies, Practices, and Technologies course consists of nine modules that include the following:

- Introduction and Theory.
- Kinetics.
- Riot Control Agents and Related Technologies.
- Maritime Vehicle Stoppers.
- Advanced Technologies.
- Emerging Technologies.
- Public Order.
- Integration of Nonlethal Weapons in Decision-Making.

1-24. Additional nonlethal weapons training material available from Penn State's Center for Community and Public Safety include the following publications available on DVD or CD:

- Overview of Kinetic NLW.
- Medical Factors and Human Effect Considerations of NLW.
- Escalation of Force Multi-media Simulations, Convoy Operations and Cordon and Search.
- Joint Nonlethal Training and Resource Manual.

Chapter 2

Conduct of Training

TRAINING STRATEGY

2-1. Training is a prerequisite for Army readiness to use NLW. The NLW training strategy is primarily organized around leader and user training. Training should be designed to give individuals an understanding of the concepts and principles of NLW and certification and familiarization on nonlethal munitions. NLW training should be integrated into all appropriate battle simulations, command post exercises, and field training exercises.

2-2. There are three objectives to consider when using NLW. The objectives include; accuracy (hitting the target in the desired location), effectiveness (are the munitions producing the desired effect on the target), and reduction of injury (are the injuries inflicted truly nonlethal). All NLW training should proceed with these objectives in mind.

USER TRAINING

2-3. User training is provided through an MMTSP designed to conduct Train-the-Trainer certification. The MMTSP is also available through the Army training Web site to ensure widest possible dissemination and rapid updating.

2-4. User-level training consists of certification and familiarization training for leaders and Soldiers on the equipment operation, nonlethal tactics, munitions, and maintenance of NLCS equipment during pre-deployment training at the unit level. Training is accomplished by the unit using the MMTSP and this TC. Units should use Soldiers assigned the ASI 2A (Nonlethal Advisor) to assist in developing a nonlethal training program that supports the units mission and to ensure the training is conducted to standard.

2-5. Once the concepts and principles of NLW employment have been mastered, nonlethal capabilities should be integrated into other training opportunities. For example, commanders should incorporate nonlethal capabilities into collective training such as cordon and search, convoy operations, checkpoint operations, detainee operations, and civil disturbance. The nonlethal equipment, tactics, and munitions described in this TC are specifically designed to support such missions. Some components are more technical and may need additional training however, most items included in the NLCS are easily mastered and relatively simple to employ.

NONLETHAL MUNITIONS TRAINING REQUIREMENTS

2-6. Commanders should ensure Soldiers assigned M203 launchers and 12 gauge shotguns receive the required training as described in DA Pamphlet 350-38.

- All units being deployed on missions which require the employment of non-lethal munitions will ensure all Soldiers/crews qualify with their assigned and/or designated weapons if they have not qualified within the past 90 days.

- Due to constraints on the availability of non-lethal munitions, this training may be done once Soldiers arrive at their deployment site. Upon notification of deployment, commanders must ensure all Soldiers who might be required to employ nonlethal munitions are properly trained in accordance with the guidance provided in this TC.

- Ninety percent of the Soldiers identified to train with non-lethal munitions will have met the individual requirements to standard with their assigned weapons.

Chapter 2

CERTIFICATION AND FAMILIARIZATION TRAINING

2-7. Certification of nonlethal munitions requires the individual to demonstrate safe handling procedures of a particular munition and firing platform and to discharge it with a reasonable degree of accuracy while completing a prescribed course of fire. Certification acknowledges that there has been a minimum standard of performance set and that the individual met the standard.

2-8. Familiarization, on the other hand, includes many of the tasks and conditions of a qualification but does not include a minimum standard. An individual demonstrates that they can perform loading, unloading, aiming, and discharging the system safely. Whether they hit the target, where they hit the target, or how many times they can repeat it is not important to the task. Accuracy and shot placement as well as fine tuning sweet spot recognition is not measured in the performance.

2-9. A qualification for nonlethal rounds is more than just hitting a target at a given distance with a given number of rounds. This is just part of the requirement. Other critical factors just as important include the aim point and the sweet spot for the round. These factors are described in the following sections.

AIMING POINT

2-10. Achieving the proper aiming point is critical for all nonlethal munitions training. Where a particular nonlethal projectile is aimed will play a great part in what effect the round will have on the target. Typically, Soldiers are taught to aim center mass. However, striking larger muscle groups such as the thigh or lower abdomen are more effective for nonlethal munitions and reduce the chance of serious injury or death. Striking the head, neck, spine, or solar plexus are targets that will result in greater risk of serious injury or death. Aiming points for specific rounds are discussed in each of the chapters on nonlethal munitions. Figure 2-1 illustrates nonlethal shot placement zones.

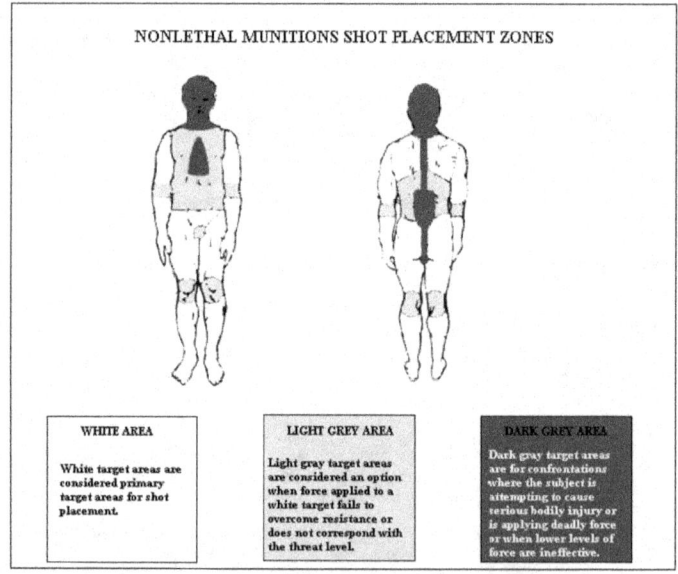

Figure 2-1. Nonlethal Munitions Shot Placement Zones

SWEET SPOT

2-11. The "sweet spot" is the finite distance between the point away from the launcher muzzle where the projectile has lost enough energy that it will not cause serious injury and the point where the projectile has lost so much energy that a hit is no longer effective and/or accurate. The sweet spot will be different for each munition, but typically it is between 10 and 55 meters.

2-12. Proper target range estimation is an essential skill for individuals using nonlethal munitions. Intentionally using a nonlethal round closer than the minimum safe distance may increase the risk of serious injury or death. Using the munition beyond the maximum effective range will result in a wasted shot or the striking of an unintended target as the round's flight path degrades as it looses its energy. Table 2-1 illustrates the effects of nonlethal munitions at various distances.

Table 2-1. Effects of Nonlethal Munitions at Distances

Type of Nonlethal Munition	Potential to Cause Serious Injury or Death	Sweet Spot	Potentially Ineffective
40MM Sponge Grenade Round, M1006	0m – 10m	10m – 50m	50m+
40MM Crowd Dispersal Round (area), M1029	0m – 10m	10m – 30m	50m+
12-GA Crowd Dispersal Round (area), M1013	0m – 10m	10m – 20m	20m+
12-GA Fin Stabilized Round, M1012	0m – 10m	10m – 20m	20m+
Modular Crowd Control Munition, M5	0m – 5m	5m – 15m	15m+
66MM Nonlethal Grenades, L96, M98, & M99	0m – 80m	80m – 100m	100m+

ACCURACY

2-13. Accuracy is a key factor that works in combination with sweet spot and aim point to achieve the desired effect in the targeted subject. Nonlethal rounds while in some cases are considered accurate, will never be as accurate as lethal rounds. They are most accurate at the beginning of their sweet spot and less accurate at the terminal end of the sweet spot. This is due to their loss of energy and the possible effects of the environment such as wind, rain, or extreme temperature on relatively slow and large rounds.

2-14. The only way to predict how a particular nonlethal munition will perform across the sweet spot area is to train with it and the appropriate firing platform. Accuracy is where the sweet spot and aim point intersect on the target to produce a desired effect.

2-15. Some nonlethal munitions are designed to be fired directly at a specific single target while others are designed to be used primarily as area target munitions. Area target munitions are multiple rounds that will deliver several rubber balls into a narrow area against several targets in a specified location.

ENGAGEMENT SKILLS TRAINER (EST) 2000

2-16. The EST 2000 is an automated computerized simulator that provides realistic weapons qualification and familiarization to a Soldier or operator. The video-based system has more than 30 high-fidelity friendly and enemy targets, 14 terrain sets, variable climatic conditions and special effects combined with over 175 scenario options. Soldiers may familiarize, and when NL munitions are not available, Soldiers may qualify and certify using the EST 2000 when authorized by the commander.

2-17. The EST 2000 allows Soldiers to fire an array of different weapons to familiarize them with the feel, sound and sight of weapons such as the M203 Grenade Launcher and the M1200 12 Gauge Shotgun. Table 2-2 and 2-3 illustrate individual NLW training currently programmed and available on the EST 2000.

Table 2-2. 40MM Sponge Grenade and 12-GA Point Round Individual Training Program

INDIVIDUAL TRAINING PROGRAM		
M203 [40MM] (M1006 Sponge Grenade) & M1200 12 Gauge (M1012 Point Round)		
Table	Distance (Meters)	Number of Target Exposures
1	7M	07
2	21M	08
3	29M	07
4	34M	05
5	49M	13

Table 2-3. 40MM Crowd Dispersal and 12-GA Crowd Dispersal Individual Training Program

INDIVIDUAL TRAINING PROGRAM		
M203 [40MM] (M1029 Crowd Dispersal) & M1200 12 Gauge (M1013 Crowd Dispersal)		
Table	Distance (Meters)	Number of Target Exposures
1	25M	04
2	25M	04
3	25M	04

2-18. The EST 2000 has 10 different collective training scenarios written into the program. The crowd control scenario allows the Soldier to respond with either lethal or nonlethal depending on the circumstances involved.

COMPOSITE RISK MANAGEMENT

2-19. Composite risk management (CRM) is the Army's primary decision-making process for identifying hazards and controlling risks across the full spectrum of Army missions, functions, operations, and activities. CRM is used to mitigate risks associated with all hazards that have the potential to injure or kill personnel, damage or destroy equipment, or otherwise impact mission effectiveness. Refer to FM 5-19.

2-20. Risk assessments are conducted to make operations safe without compromising the level of realistic training. Commanders continuously assess the risk of training conditions to prevent injury to Soldiers and loss of equipment. Risk assessments should consider the level of proficiency of the Soldiers and trainers as well as environmental factors such as weather and terrain. Training enhancers, such as role-players, demonstration fire, blank ammunition, simulators, smoke, and other pyrotechnics, are fully integrated into NLW training to achieve the training objective. Soldiers are required to wear ear protection, protective vests, and eye protection in order to protect against noise and debris.

2-21. The commander is the safety officer and is responsible for ensuring that his/her unit, both leaders and Soldiers, comply with safety regulations and the unit tactical standard operating procedure (SOP). All leaders must:

- Use mission, enemy, terrain and weather, troops and support available, time available, and civil considerations (METT-TC) factors to identify risks.
- Assess possible losses and their costs.
- Select and develop risk-reduction measures.
- Implement controls by integrating them into plans and order, SOP, training performance standards, and rehearsals.
- Never shoot nonlethal munitions at human targets during training.
- Supervise and enforce risk reduction measures and safety standards at all times.

RECOMMENDED RANGE PROCEDURES

Note: Before conducting a live-fire range, ensure you adhere to the procedures that support Army regulations, local range regulations, and established unit training SOPs.

2-22. Before beginning a live-fire exercise, all personnel must receive an orientation on range operations. The orientation should outline the procedures for conducting the exercise to include the duties of the non-firing orders. To provide a safe and efficient range operation and effective instruction, the following is an example of personnel and duties that may be required.

- Officer in Charge. The Officer in Charge (OIC) is responsible for the overall operation of the range before, during, and after live firing.
- Range Safety Officer. The range safety officer (RSO) is responsible for the safe operation of the range to include conducting a safety orientation before each scheduled live-fire exercise. He ensures that all personnel comply with the safety regulations and procedures prescribed for the conduct of a live-fire exercise. He ensures that each weapon is cleared before a firer leaves the firing line. The RSO will not be assigned any other duties.
- Non-Commissioned Officer in Charge. The Non-Commissioned Officer in Charge (NCOIC) assists the OIC and safety officer, as required. For example, the NCOIC will supervise enlisted personnel who are supporting the live-fire exercise.
- Ammunition Detail. This detail is composed of one or more ammunition handlers whose responsibilities are to break down, issue, receive, account for, and safeguard live ammunition. The detail also collects expended ammunition casings and other residue.
- Unit Armorer. The unit armorer repairs the weapons to include replacing parts, as required.
- Lane Safeties. One assistant instructor (AI) is assigned for each one to two firing points. Each lane safety ensures that all firers observe safety regulations and procedures and assists firers having problems.
- Medical Personnel. They provide emergency medical treatment and medical evacuation support as required by local policy.
- Control Tower Operator. The tower operator gives the fire commands.
- Maintenance Detail. This detail performs minor maintenance on the target-holding mechanisms.
- Communications. At least two (2) means of communication are needed to maintain continuous communications with Range Control. Local policy will dictate which forms of communication are allowed.

2-23. Factors to consider when conducting a nonlethal live fire exercise:

- Develop a Range Safety SOP if one does not exist.
- Instructor to student ratio is generally 1:10.
- Accidents happen from lack of supervision. Supervise continuously.
- Nonlethal munitions can kill or seriously injure. Use the same safety standards used for lethal munitions.
- Train as you fight.
- Always have medical personnel present.

2-24. General Range Safety Considerations: These range rules apply to any range situation and should be followed and enforced at all times.

- Hearing protection, eye protection, and appropriate headgear will be worn on and in the vicinity of the firing line.
- All firing activities will be done on command.
- Do not handle or pick up any weapon until told to do so.
- Do not handle a weapon while anyone is down range.
- No one goes forward of the firing line unless directed to do so by Range Cadre.
- Keep weapons pointed down range at all times.
- Keep weapons on safe when not actually firing.
- Keep trigger finger off the trigger when not actually firing.
- Follow proper weapon jam or malfunction procedures.

RULES OF ENGAGEMENT

2-25. ROE are the directives established by higher Headquarters that delineate the circumstances and limitations under which Soldiers will initiate and/or continue engagement with belligerent forces. ROE may reflect the law of armed conflict and operational considerations but are primarily concerned with the restraints on the use of force. ROE are the primary means by which commanders convey legal, political, diplomatic, and military guidance to Soldiers. Leaders at every level must train Soldiers carefully and thoroughly concerning ROE and laws that govern armed conflict before deployment. During the conduct of operation, leaders continue to train Soldier and stress firm, determined, and impartial execution of ROE.

RULES FOR THE USE OF FORCE

2-26. Since NLW are often considered for civil support operations commanders should ensure Soldiers participating in these operations are trained on Rules for the Use of Force (RUF). RUF provide the operational guidance and establish fundamental policies and procedures governing actions taken by DoD forces performing civil support missions within the U.S. and its territories. RUF apply to land-based homeland defense missions occurring within the U.S. and its installations within or outside the U.S. and its territories. For Army law enforcement and security personnel *AR 190-14* provides policy for RUF.

FORCE CONTINUUM TRAINING

2-27. The force continuum concept provides a guide for the use of force that must be reasonable or proportional in intensity, duration, and magnitude based on the totality of circumstances. This continuum concept states that there is a wide range of possible actions, ranging from presence and verbal commands to application of deadly force that may be used to counter resistance or threat in any particular situation. Situations vary, and the threat level can rise and fall several times based on the actions of both the Soldier and the individual or groups involved. The purpose of a force continuum concept is to serve as a guide and a graphic training aid to assist Soldiers in use of force decision making consistent with ROE, commander's intent, and the force options available to the Soldier.

2-28. This stair-stepped approach shows the action and reaction correlation between the Soldier and the individual or group. Typically, there are five levels of subject behavior and Soldier response:

1. Compliant: If the subject is compliant, the Soldier should remain professional and use cooperative controls.
2. Passive Resistance: If the subject has become passive resistant, the Soldier may assume a tactical posture and consider the use of contact controls.

3. Active Resistance: If the subject has become assaultive, the Soldier may reach a threshold and will have to consider the use of compliance techniques.
4. Assaultive: Once the subject has become assaultive and does not respond to compliance techniques, the Soldier may have to employ defensive tactics which may include the use of nonlethal munitions.
5. Serious/Grievous Bodily Harm or Death: Once the subject has become assaultive to the point of bringing serious bodily harm or death, the Soldier may have to employ deadly force.

2-29. NLW training must include the use of force decision-making process, how to leverage nonlethal capabilities into mission planning, tactical nonlethal considerations, and how to incorporate lethal and nonlethal force options. Soldiers must be taught these skills through situational training exercises where they are required to constantly travel up and down the force continuum stair steps. Figure 2-2 illustrates the five levels of subject behaviors and Soldier response.

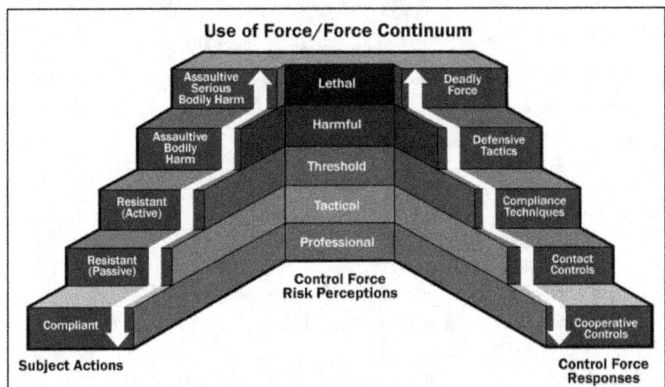

Figure 2-2. Force Continuum

ESCALATION OF FORCE

2-30. The following information is derived from the Center for Army Lessons Learned (CALL) Escalation of Force Handbook (No. 07-21). The purpose of escalation of force (EOF) training is to improve leader and Soldier awareness when planning, preparing, and executing EOF methods to prevent unnecessary deaths of Soldiers and civilians. The desired end state is for Soldiers and leaders at all levels to understand and apply EOF procedures in order to lengthen response time and prevent unnecessary lethal force engagements. The following EOF-related tasks should be integrated into NLW training:

- Soldiers at every level must understand EOF procedures before, during, and after an EOF event to include the possible second and third order effects and/or the strategic impact of EOF incidents, which can lead to misperceptions by the local populace.
- Soldiers must continually train and rehearse EOF procedures at home station, during mobilization training and mission rehearsal exercises, and while deployed.
- All Soldiers must understand EOF in order to prevent hesitation or second-guessing when making a decision to apply force.
- Units should be resourced with the correct protection equipment in order to increase reaction time, reduce unnecessary EOF incidents resulting in the use of lethal force, and reduce casualties.

2-31. When time and circumstances permit, the forces committing hostile acts or demonstrating hostile intent should be warned and given the opportunity to withdraw or cease threatening actions. The opportunity to de-escalate a situation may result from the mere presence or drawing of a NLW.

2-32. If these important EOF-related tasks are successfully integrated into training, and Soldiers and leaders adhere to the training principles, unnecessary lethal force engagements can be reduced or eliminated. Adhering to EOF-trained principles also supports the goal of enhancing the partnership with local governments by reinforcing a positive image of coalition forces. Key benefits of EOF training include the following:

- EOF-trained Soldiers are able to make the right decision on what is a legitimate target.
- EOF training lengthens the preparation time to positively identify targets and apply force commensurate with the level of threat.
- EOF never limits a Soldier's inherent right to self-defense.

DOCTRINAL GUIDANCE FOR NONLETHAL WEAPONS

2-33. Doctrinal guidance for the employment of NLW/capabilities can be found throughout the Doctrine, Organization, Training, Materiel, Leadership and Education, Personnel, and Facilities (DOTMLPF) domains. Tactics, Techniques, and Procedures (TTP) for the tactical employment of NLW are described in FM 3-22.40. FM 3-22.40 is an "all service" manual that was developed by the Air Land Sea Application Center. In addition, several 3-19 series field manuals describe NLW employment during specific military police operations. Figure 2-3 highlights doctrinal and training resources that are essential for Soldiers, leaders, and staffs responsible for planning and conducting nonlethal training.

Doctrinal Publications
• FM 3-22.40, *Multi-Service TTP for the Tactical Employment of Nonlethal Weapons*
• FM 3-07, *Stability Operations*
• FM 3-19, *Military Police Operations*
• FM 3-19.4, *Military Police Combat Leaders' Handbook*
• FM 3-19.11, *Military Police Special Reaction Teams*
• FM 3-19.15, *Civil Disturbance Operations*
• FM 3-19.40, *Military Police Internment and Resettlement Operations*
• FM 3-63.1, *Detainee Operations*
• FM 4-01.45, *Multi-Service Tactics, Techniques, and Procedures for Tactical Convoy Operations*
• TRADOC Pamphlet 525-99, *Concept for Nonlethal Capabilities in Army Operations*
• DA Pamphlet 350-38, *Standards in Training Commission* (STRAC)
• CALL Handbook, No. 07-21, *Escalation of Force*, August 2007
Training Resources
• Inter-Service Nonlethal Individual Weapons Instructor Course (INIWIC)
• Nonlethal Multi-Media Training Support Package
• Penn State Online Nonlethal Weapons: Policies, Practices, and Technologies Certificate Program
• Engagement Skills Trainer (EST) 2000

Figure 2-3. Doctrinal and Training Resources

Chapter 3

Employ 12-Gauge Point and Area Target Munitions

OVERVIEW

3-1. Nonlethal 12-gauge munitions are used to control hostile crowds by temporarily incapacitating without causing life-threatening consequences. This gives commanders the option to apply nonlethal force as a first line of defense where appropriate.

3-2. The firing platform for the Army 12-gauge nonlethal munitions is the 12-gauge pump action shotgun such as the Mossberg 500. Other models that may be used include the Winchester M1200 and the Remington 870. In addition to firing from a standard 12 gauge pump shotgun, both the M1012 and M1013 can be fired from the 12 gauge Modular Accessory Shotgun System (MASS). MASS is a lightweight accessory shotgun system that attaches underneath the barrel of the M4 and M16 rifle. It provides the capability to fire lethal, nonlethal, and door breaching 12 gauge rounds. It can be zeroed to the sighting system of the host weapon. MASS provides the lethality equivalent of a stand-alone 12 gauge shotgun.

M1012 12-GAUGE NONLETHAL POINT ROUND SPECIFICATIONS

3-3. The M1012 point round is a nonlethal, low hazard, non-shrapnel producing device fired from a 12-gauge pump-action shotgun. It is a standard 12-gauge shotgun shell, cylindrical in shape with a three-quarter-inch diameter and is 2.5 inches long. The payload is a molded rubber fin-stabilized projectile fired at a muzzle velocity of about 500 feet per second. It can be fired from both the 2.75-inch and 3-inch chamber of standard military pump-action shotguns. The M1012 national stock number (NSN) is 1305-01-470-2405.

3-4. The round's sweet spot is between 10 to 20 meters. Closer than 10 meters the round may create lethal or near lethal injuries to the target. Beyond 20 meters the round will be inaccurate and ineffective. While distance to the target is critical, round placement is also a critical factor. Head and neck shots are not acceptable.

3-5. Figure 3-1 provides a cut-away view of the M1012 12-gauge Point Round.

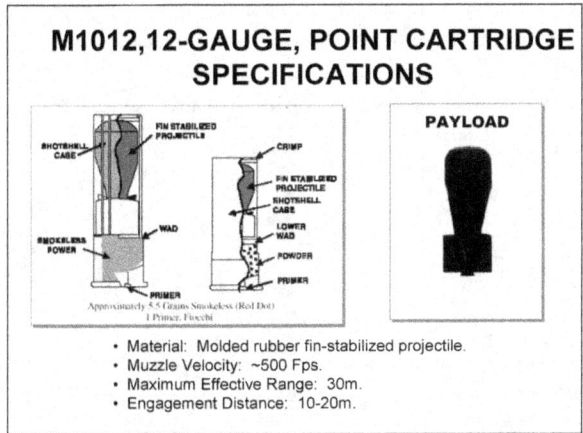

Figure 3-1. Cut-away View of the M1012 12-Gauge Point Cartridge

> **WARNING:** All nonlethal rounds are subject to "bounce back". Firers and range personnel must use caution.

M1012 EMPLOYMENT CONSIDERATIONS

3-6. The M1012 point control cartridge gives the Soldier the capability to stun individuals without penetration by delivering a strong blow to the body. The munition fills a wide range of possible NLW applications. In crowd control, the M1012 point round provides a multi-shot nonlethal capability that can be used to break contact, enforce a buffer zone (stand-off distance) with a violent crowd, or stun an individual threat.

3-7. The cartridge is designed to inflict nonlethal blunt trauma to a single targeted individual thereby, increasing flexibility in the application of force during military operations. It is intended to confuse, disorient, and distract a person who may be a potential threat.

M1012 CERTIFICATION STANDARDS

3-8. Each firer will conduct Tables 1 through 5 on the EST 2000 system as part of preliminary marksmanship instruction prior to firing the certification Table. Each firer will have five rounds for certification and five rounds for annual training in one of the collective tasks listed in Table 3-1. Use the following guidance for M1012 point cartridge certification.

- Range: Set target at 15 meters.
- Target: Standard E-type silhouette.
- Time: 60 seconds.
- Shooting position: Standing unless typical mission requires an alternate position such as kneeling.
- Minimum score: Four hits at the designated target area.

3-9. Instructor or designee will be responsible for counting hits on target. Hits can easily be counted visually as the round strikes the target.

Table 3-1. Collective Tasks for Annual Nonlethal Weapons Training

TASK NUMBER	TYPE OF TASK	TASK TITLE
19-2-3512	Collective	Respond to Emergency Situations in an Internment/Resettlement (I/R) Facility and/or Camp
19-3-2406	Collective	Roadblocks and Checkpoints Operations
19-2-4003	Collective	Conduct Company Level Civil Disturbance Control Operations
19-2-2014	Collective	Supervise In-Transit Security

M1012 SIGHTING

3-10. The M1012 point round is designed to engage targeted individuals at a range of 10 to 20 meters. Do not engage if target is closer than 10 meters unless lethal force is justified. Serious injury or death of targeted individual may result.

3-11. With the shooting eye, align top center of receiver assembly with the front sight centered on target mass. Concentrate to keep sight on target. Always target the center of mass and not the head or face of an individual. Do not "skip fire" the projectiles at the targeted individual. Accuracy and range will be decreased. Tumbling of the projectile may result in serious injury to the targeted individual.

3-12. Figure 3-2 shows the proper aiming point for targets 10 to 20 meters away.

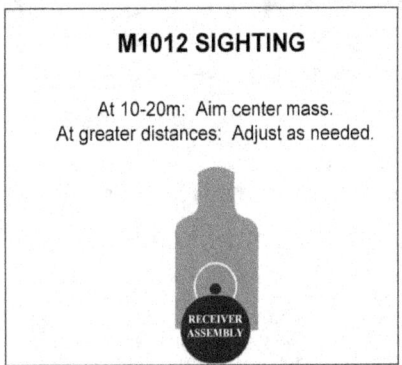

Figure 3-2. M1012 Sighting at 10-20 Meter Targets

M1013 12-GAUGE CROWD DISPERSAL CARTRIDGE SPECIFICATIONS

3-13. The M1013 Crowd Dispersal Cartridge is a standard 12-gauge shotgun shell, cylindrical in shape with a three-quarter-inch diameter and 2.5 inches length. The payload consists of 18 small molded rubber pellets fired at a muzzle velocity of about 900 feet per second. It can be fired from both the 2.75-inch and 3-inch chamber of standard military shot guns. The M1013 NSN is 1305-01-470-2139.

3-14. The rounds sweet spot is 10 to 20 meters. Closer than 10 meters it may have lethal effects. Farther than 20 meters it will be ineffective. Placement of round is a critical factor. Head and neck shots are not acceptable unless lethal force is justified. Figure 3-3 is a cut-away view of the M1013 12-gauge Crowd Dispersal Cartridge.

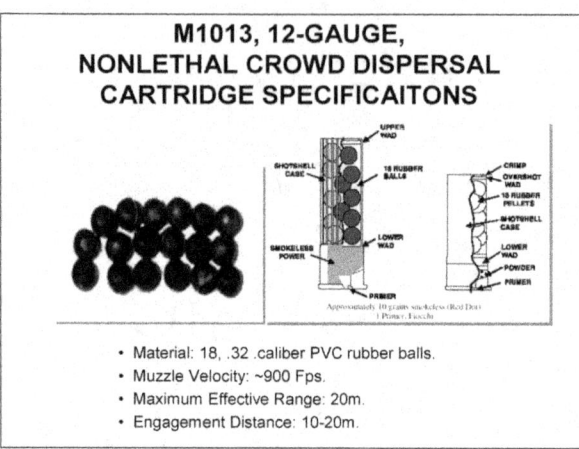

Figure 3-3. Cut-away View of the M1013 12-Gauge Crowd Dispersal Cartridge

M1013 EMPLOYMENT CONSIDERATIONS

3-15. The M1013 Crowd Dispersal Cartridge is an area target round that gives the Soldier the capability to stun/deter two (2) to three (3) people without penetration by delivering a strong blow to the body. These munitions fill a wide range of possible NLW applications. In crowd control, the M1013 Crowd Dispersal Cartridge gives a Soldier a multi-shot nonlethal capability that can be used to break contact, enforce a buffer zone (stand-off distance) with a violent crowd, or clear an area of multiple threat targets.

3-16. The M1013 Crowd Dispersal Cartridge is intended to confuse, disorient, and distract a person who may be a potential threat to friendly force personnel. The M1013 Crowd Dispersal Cartridge is a nonlethal, low hazard, non-shrapnel-producing device fired from a 12-gauge pump-action shotgun. The cartridge is designed to inflict nonlethal blunt trauma to multiple threat targets with one shot.

M1013 FAMILIARIZATION STANDARDS

3-17. Each firer will conduct Tables 1 through 5 on the EST 2000 system as part of preliminary marksmanship instruction prior to firing familiarization fire. Each firer will have two rounds for familiarization. Use the following guidance for M1013 round qualification:

- Range: Target set at 15 meters.
- Target: Standard E-type silhouette.
- Time : 25 seconds.
- Shooting position : Standing unless typical mission requires an alternate position.

3-18. Firers will only familiarization with multiple projectile munitions due to the time required for counting, scoring, and marking targets.

M1013 SIGHTING

3-19. The M1013 Crowd Dispersal Cartridge is designed to engage a group of targeted individuals at a range of 10 to 20 meters. With the shooting eye, align top center of receiver assembly with the front sight centered on target mass. Concentrate to keep sight on target. Always target the center of mass and not the head or face of an individual in the crowd. Do not engage if target is closer than 10 meters unless lethal force is justified. Serious injury or death of targeted individual may result. Figure 3-4 illustrates the proper aiming point for targets 10 to 20 meters.

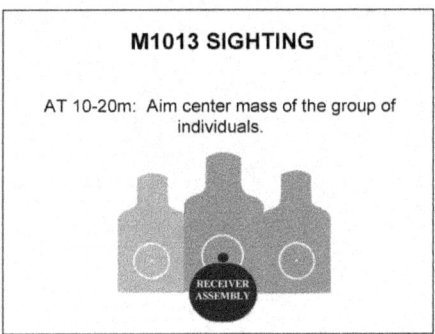

Figure 3-4. M1013 Sighting at 10 to 20 Meters

This page intentionally left blank.

Chapter 4

Employ 40 Millimeter Nonlethal Munitions

OVERVIEW

4-1. The U.S. Army currently fields the M1006 point target round better known as the "Sponge Grenade" and the M1029 Crowd Dispersal Cartridge. Each is intended for close quarter engagement and can engage 3 or more targets at once.

4-2. The firing platform for the U.S. Army 40mm nonlethal munitions is the M203 grenade launcher mounted on M16 or M4 series weapons. Only individuals assigned the M203 should complete the annual certification requirements.

4-3. Both the M1006 and M1029 can be fired from the M32 Launcher however they will not cycle the M32 multi-shot launcher. These rounds can not be fired from the MK19 Machine Grenade Launcher.

M1006 SPONGE GRENADE SPECIFICATIONS

4-4. The M1006 Sponge Grenade is designed with a foam rubber nose affixed to a carrier as the payload. The foam rubber nose is why it is commonly referred to as the "Sponge Grenade." It is a direct-fire, low-hazard, non-shrapnel-producing projectile. The M1006 NSN is 1310-01-452-1190.

4-5. The M1006 has a muzzle velocity of 265 feet-per-second with a terminal velocity of 200 feet-per-second at 50 meters. The sweet spot for the Army Sponge Grenade is 10 to 50 meters.

4-6. The payload is installed and secured into a high density, plastic-cartridge case. An impulse cartridge installed in the rear of the cartridge case develops the propulsive energy. The impulse cartridge has the primer. Like an M781 training/practice cartridge, no propellant materiel is in the plastic cartridge case. Figure 4-1 is a cut-away view of the M1006 Sponge Grenade.

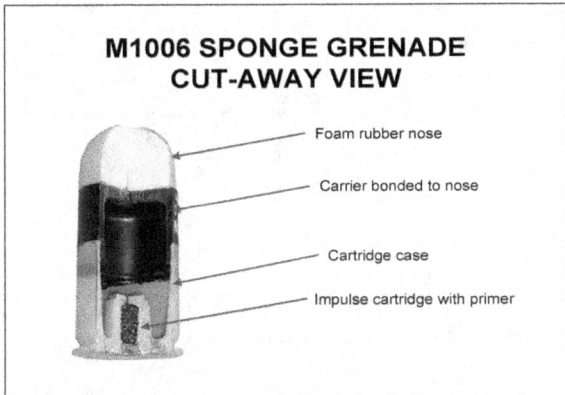

Figure 4-1. M1006 Sponge Grenade Cut-Away View

M1006 EMPLOYMENT CONSIDERATIONS

4-7. The M1006 Sponge Grenade is a 40mm nonlethal point round that gives the Soldier the capability to stun individuals without penetration by delivering a strong blow to the body. This munition fills a wide range of possible NLW applications.

4-8. In crowd control, M1006 Sponge Grenades provides a nonlethal capability that can be used to break contact, enforce a buffer zone (standoff distance) with a violent crowd, or stun an individual threat for possible detention. The M1006 Sponge Grenade can also provide detention facilities with similar capabilities. This round is used to confuse, disorient, dissuade, and gain control of hostile individuals in a wide range of military, law enforcement, or civil unrest operations. It provides the commander an additional tool for controlling unruly subjects without relying solely on the use of deadly force.

4-9. When using the M1006 Sponge Grenade the following employment procedures should be considered:

- Do not engage individuals closer than 10 meters (except in situations where deadly force is justified).
- Do not fire at individuals above chest level.
- Be aware of "bounce-back." If projectile is fired at a hard object or wall within 20 meters of the shooter.
- Do not ricochet the round off the ground. This could cause the projectile to rotate and strike the targeted individual with the hard plastic carrier.

M1006 CERTIFICATION STANDARDS

4-10. Each M203 firer will have five rounds for familiarization and five rounds for qualification. Use the following guidance for M1006 round qualification:

- Range: Targets set at 10, 20, and 30 meters.
- Target: Standard E-type silhouette.
- Time: 60 seconds.
- Shooting position: Standing unless mission requires alternate position such as kneeling.
- Minimum score: Four hits at the designated target area. Rounds which hit the head, neck, or sternum will be considered a miss. Mark record as pass or fail.

4-11. Instructors or designee will be responsible for counting hits on target. Hits can be counted visually as the hits are made on the target.

M1006 SIGHTING

4-12. When aiming, use the 50-meter notch of the leaf sight on the M203 Grenade Launcher and the front sight of the M4 Carbine or M16 series rifle for aiming at targets when firing the M1006 Sponge Grenade cartridge. The 50-meter notch is identified by a red line on the leaf sight. When the targeted individual is 50 meters from the shooter, align the sights on the weapons with the center of mass on the target.

4-13. The operator must sight/aim below the intended target impact point for targets that are 45 meters or less from the shooter. Align the sights approximately 24 inches below the center of mass for targets that are between 10 and 45 meters away from the shooter. This is done to compensate for projectile rise when firing at targets less than 50 meters from the shooter. Adjust your aim point as necessary.

4-14. Figures 4-2 and 4-3 illustrate the M1006 aiming points at 50 meters and 10 to 45 meters.

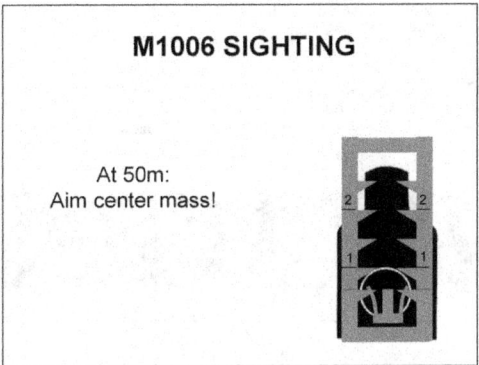

Figure 4-2. M1006 Sighting at 50 Meters

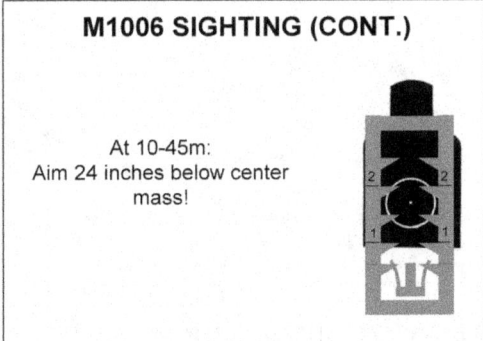

Figure 4-3. M1006 Sighting at 10 to 45 Meters

M1029 40MM NONLETHAL CROWD DISPERSAL CARTRIDGE SPECIFICATIONS

4-15. The M1029 40mm round is an aluminum cartridge of similar proportions to standard 40mm illuminant cartridges but with no separate cartridge case. A fiberboard sleeve and plastic cover contain the internal nonlethal payload of 48, 0.48 caliber rubber balls that are designed to cause nonlethal blunt trauma to a targeted group of individuals upon impact. The M1029 NSN is 1310-01-475-0628.

4-16. When this round is fired from the M203 grenade launcher, it acts much like a large shotgun shell. The payload spreads out from the barrel to cover an area equal to five standard E-type silhouettes standing side-by-side at a range of 30 meters.

4-17. The M1029 has a sweet spot of between 10 and 30 meters. It may have lethal effects if discharged into a target inside of 10 meters and is ineffective and not accurate after 30 meters. While distance to the target is critical, round placement is equally important. This round should not be fired at the upper chest, neck, or head. Placing a round in an eye socket can prove fatal at any "safe" range while a thigh hit would rarely, if ever, be fatal at any distance.

4-18. The M1029 Crowd Dispersal Cartridge is an area fire, nonlethal, low hazard, non-shrapnel-producing munition that is fired from the M203 grenade launcher mounted on either the M4 carbine or M16 series rifle. This round cannot be fired from the MK-19 Machine Grenade Launcher. Figure 4-4 is a cut-away view of the M1029 Cartridge.

> **WARNING:** All nonlethal rounds are subject to "bounce back". Firers and range personnel must use caution.

M1029, 40 MM
NONLETHAL CROWD DISPERSAL CARTRIDGE
SPECIFICATIONS

- Payload: 48 Ea 0.48 Caliber Rubber Balls.
- Muzzle Velocity: ~ 375 Feet/Second.
- Terminal Velocity At 30m: ~ 200 Fps.
- Minimum Engagement Distance: 10m.
- Maximum Effective Range: 30m.

WARNING: The M1029 Cannot Be Fired From The MK19 MGL.

Figure 4-4. Cut-Away View of the M1029

M1029 EMPLOYMENT CONSIDERATIONS

4-19. The M1029 Crowd Dispersal Cartridge gives the Soldier the capability to stun/deter two (2) to three (3) people without penetration by delivering a strong blow to the body. These munitions fill a wide range of possible NLW applications. In crowd control, the M1029 Crowd Dispersal Cartridge delivers a multi-shot nonlethal capability that can be used to break contact, enforce a buffer zone (standoff distance) with a violent crowd, or clear an area of multiple threat targets. It can also provide detention facilities with similar capabilities. It is intended to confuse, disorient, or momentarily distract potential threat personnel.

4-20. This device can be used by military personnel in crowd control operations, in hostage rescue situations, and in the capture of criminals, terrorists, or other adversaries. It provides commanders a nonlethal capability to increase the flexibility in the application of force during military operations.

4-21. The M1029 is subject to wind and gravity effects at longer ranges. When firing in windy conditions (approximately 10 mph) some hold-off will be required when firing these items. As range increases, the rounds will drop toward the ground. If instinct shooting, focus your eyes on the center of mass of the target, not at the target's head. Adjust aiming point as necessary for condition. When using the M1029 Crowd Dispersal Cartridge:

- Do not engage individuals closer than 10 meters unless deadly force is justified.

- The payload will spread to hit at least five (5) individuals standing side-by-side at 30 meters.
- Be aware of possible bounce-back of payload when firing at hard targets within 20 meters of shooter.

M1029 FAMILIARIZATION STANDARDS

4-22. Each M203 firer will have two rounds for familiarization. Use the following guidance for M1029 round familiarization:

- Range: Set target at 30 meters.
- Target: Standard E-type silhouette set up in groups of three abreast per shooter.
- Time: 30 seconds.
- Shooting position: Standing unless typical mission requires an alternate position such as kneeling.

M1029 SIGHTING

4-23. The iron sights of the M16 rifle or M4 carbine are used for aiming when using the M1029 Crowd Dispersal Cartridge. The aiming point on the targeted hostile group is center mass. Select an individual in the middle of the crowd. If targeted group is more than 15 meters away, aim at the center of the torso (chest high). If the targeted group is 15 meters or closer, aim at the waist or legs to reduce head hits.

4-24. Figures 4-5 and 4-6 illustrate proper M1029 aiming points at 15 to 30 meters and 10 to 15 meters.

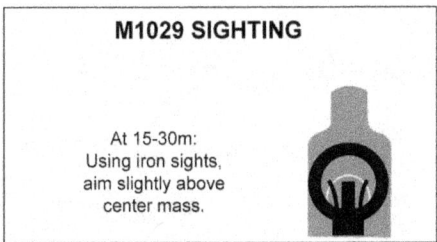

Figure 4-5. M1029 Sighting at 15 to 30 Meters

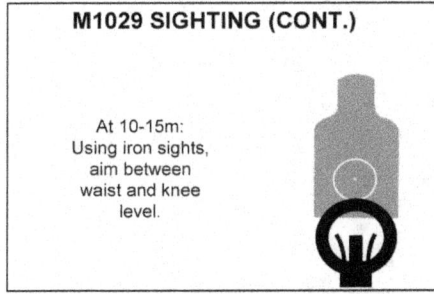

Figure 4-6. M1029 Sighting at 10 to 15 Meters

This page intentionally left blank.

Chapter 5

Employ Nonlethal Grenades

OVERVIEW

5-1. The U.S. Army currently fields two hand-tossed low hazard grenades. The M84 Stun Hand Grenade is a non-fragmenting grenade designed to temporarily distract a potential threat and the M104 Sting Ball Grenade is designed to create a diversion and temporarily incapacitate. The M104, which is typically thrown by hand, can also be launched from a device attached to a standard 12-gauge shotgun.

M84 STUN HAND GRENADE SPECIFICATIONS

5-2. The M84 Stun Hand Grenade is a hand-tossed, nonlethal, low-hazard, non-fragmenting device that weighs 15 ounces. The M84 Stun Hand Grenade will produce the following effects:

- Intense light (flash) over 1 million candle-power.
- Sound (bang) ranging from 170–180 decibels at five (5) feet.

5-3. The user must wear approved single hearing protection when employing the grenade during training in the event of a premature functioning or bounce back when tossed. Personnel within 25 feet of a functioning M84 Stun Hand Grenade must also wear single hearing protection. Use of approved hearing protection during actual operations is also recommended.

5-4. Figure 5-1 shows the M84 Stun Grenade.

Figure 5-1. M84 Stun Grenade

5-5. The grenade body is a hollow steel cylinder 5.25 inches long and 1.73 inches in diameter with hexagonal top and bottom. Twelve vent holes in the cylindrical portion of the body allow the release of energy from the payload when initiated. A modified M201A1 fuse is secured in a threaded opening in the top of the body. The fuse consists of primary and secondary safety pins, a safety lever (spoon), a spring-loaded striker, and a percussion primer and metal oxidant mix. The pyrotechnic charge is a magnesium and potassium perchlorate mix. Within its operational bounds, the blast overpressure or intense flash of this device will not produce any long-term effects on vision or hearing. However, protective gloves should be worn. Personnel laying on top of or adjacent to these devices may experience serious injury or death.

5-6. The M84 Stun Hand Grenade is easily identifiable as a nonlethal device by the hexagonal shape at the top and bottom of the body, a nonlethal green band around the body, and a brown band on the spoon. The M84 Stun

Hand Grenade also has a round primary safety pin that is removed immediately prior to tossing the grenade and a triangular shaped secondary safety pin that precludes inadvertent initiation in transportation, storage, and handling. Removing the safety pins requires a pull ring force between 15 to 40 pounds. The M84 NSN is 1330-01-459-8141.

M84 EMPLOYMENT CONSIDERATIONS

5-7. The M84 Stun Hand Grenade is a non-fragmentation, nonlethal, one-time-use, "flash and bang" grenade. It has been successfully deployed in many situations by U.S. Forces to confuse, disorient, or momentarily distract a potential threat in forced entry scenarios as well as selected urban operations or crowd control operations.

5-8. The device is designed to be thrown into a room (through an open door, breached glass window, or other opening) where it delivers a loud bang and bright flash sufficient to temporarily disorient the occupants. The M84 Stun Hand Grenade is used when minimum force is necessary by tactical and non-tactical forces when performing missions such as hostage rescue or capture of criminals, terrorists, or other adversaries.

5-9. When preparing to employ the M84 Stun Hand Grenade, the user must identify an existing opening in the structure where the grenade can be tossed. If no opening exists, the user may have to create one. The user must then toss the grenade through the opening and take cover from effects. The M84 Stun Hand Grenade functions 1–2.3 seconds after spoon released. **DO NOT COOK-OFF.**

5-10. The M102 Reloadable Stun Practice Grenade is the reloadable cost effective trainer for the M84. It offers realistic characteristics of the M84 to assist in the training of stun grenades without using actual grenades.

M84 SAFETY CONSIDERATIONS

5-11. Ensure all grenades function properly. If a grenade has been identified as a "dud," cover it with a wet cloth or towel to minimize its effects in the event that it does detonate and contact EOD for proper disposal. Activation of the M84 Stun Hand Grenade should not ignite paper or cloth. However, other hazards such as volatile fumes in the space where the grenade will detonate should be considered prior to tossing it into a closed structure.

5-12. One of the byproducts generated by a functioning stun grenade is hydrogen chloride. A one-time use of the grenade in an enclosed room (approximately 20 feet by 20 feet by eight (8) feet) will not expose an individual to an unacceptable health risk. However, during training exercises, enclosed areas where stun grenades are used should be well ventilated to preclude buildup of hazardous levels of hydrogen chloride.

5-13. Trainers should have approved fire suppression devices available when deploying stun grenades.

5-14. Human target participation should not be permitted during training, even if hearing protection is provided. In addition to a potential for damage to hearing, there is also a risk of personnel injury that is not warranted during a training exercise. Possible injuries could include:

- Being struck by the grenade when tossed into a confined space.
- Grenade jumping up to 20 feet at time of detonation.
- Damage to eyes from plastic particles from the payload when detonated.
- Flash burn if individual is close to grenade when detonated.

5-15. M84 Nonlethal Stun Hand Grenades should always be transported in the original shipping configuration to preclude inadvertent removal of safety pins. Always ensure that both safety pins are installed until ready to employ the grenade. The secondary pin is provided for safety during transportation and handling and is identified by a triangular shaped pull ring. It should only be removed when you have unpackaged the grenades and are responding to a situation.

ARM M84 STUN HAND GRENADE

5-16. When removing the primary safety pin, ensure you grasp the grenade and spoon tightly. Once the primary pin is removed from the grenade and your hand lets go of the grenade spoon, the fuse will function within 1–2.3 seconds. Use the following method to employ the grenade:

- Right-handed thrower:
    1. Grasp the grenade firmly by wrapping your thumb around the spoon.
    2. Remove secondary safety pin.
    3. Remove primary safety pin.

5-17. Figure 5-2 illustrates right-handed procedure.

ARM M84 STUN HAND GRENADE (RIGHT HANDED)

RIGHT HANDED SOLDIERS:

- Grasp the grenade firmly.
- Remove secondary safety pin.
- Remove primary safety pin.

WARNING: Do not loosen your grip on the spoon once the safety pins are removed!

Figure 5-2. M84 Stun Grenade Right Handed Procedure

- Left-handed thrower:
    1. Hold the grenade with the fuse assembly pointed toward the ground.
    2. Grasp the grenade firmly by wrapping your thumb around the spoon.
    3. Remove the secondary safety pin.
    4. Remove the primary safety pin.

Chapter 5

5-18. Figure 5-3 shows left handed procedure.

**M84 STUN HAND GRENADE
(LEFT HANDED)**

LEFT HANDED SOLDIERS:

- Hold the grenade with the fuze assembly pointed towards the ground.
- Grasp the grenade firmly.
- Remove secondary safety pin.
- Remove primary safety pin.

WARNING: Do not loosen your grip on the spoon once the safety pin s are removed!

Figure 5-3. M84 Stun Grenade Left Handed Procedure

M104 STING BALL GRENADE SPECIFICATIONS

5-19. The M104 Sting Ball Grenade is primarily a hand-thrown nonlethal grenade. It is a one-time use, low hazard device that provides a diversion and causes temporary incapacitation by dispensing a payload of 100 each 0.25 inch diameter rubber balls in a 360 degree radius at a distance of 15-20 meters.

5-20. The effective radius is 2 to 3 meters, with an effective range of 15 to 20 meters, depending on whether an air-burst or ground detonation is used. The rubber balls leave the grenade at about 700 feet per second. Figure 5-4 shows a cut-away view of the M104 Sting Ball Grenade.

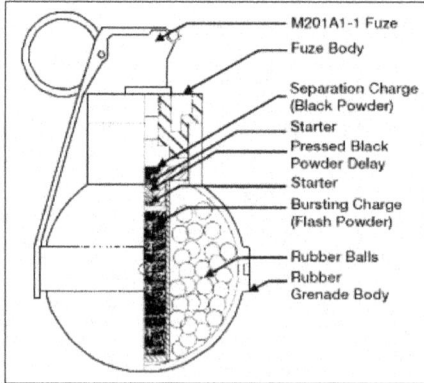

Figure 5-4. Cut-Away View of the M104 Sting Ball Grenade

M104 EMPLOYMENT CONSIDERATIONS

5-21. The M104 Sting Ball Grenade is a hand tossed grenade with two primary employment considerations. Outdoor employment requires the thrower to toss the grenade toward the target far enough away so the thrower does not feel the effects (minimum of four meters). For employing into a building or room the grenade should be tossed through an opening and the thrower should take cover from the effects.

5-22. The M104 also gives Soldiers engaged in patrolling, convoy, or crowd control a non-lethal capability to break contact, enforce a buffer zone (stand-off distance) with a violent crowd, or clear an area of a violent crowd. It operates like a standard hand grenade, detonating a few seconds after the spoon is released.

M104 SAFETY CONSIDERATIONS

5-23. The M104 has a minimum safe distance of four meters. Safety considerations for the M104 include the following:

- Handle ammunition items carefully.
- Use appropriate personal protective equipment.
- Keep grenades in original packaging prior to use.
- Safety pins installed prior to unpacking.
- Do not attempt to toss grenades through glass.
- Human targets are not allowed during training.
- Ensure enclosed areas are well ventilated.
- Maintain a tight grip around grenade body and spoon when removing safety pin.
- Retain Safety pin until grenade is used.
- DO NOT PRACTICE "COOK OFF."

ARM M104 STING BALL GRENADE

5-24. The M104 is armed the same as the M84 Stun Grenade. When removing the primary safety pin, ensure you grasp the grenade and spoon tightly. Once the primary pin is removed from the grenade and your hand lets go of the grenade spoon, the fuse will function within approximately 2.6 seconds.

5-25. Right and left handed griping procedures are the same as the M84 Stun Grenade.

This page intentionally left blank.

Chapter 6

Employ Modular Crowd Control Munition

OVERVIEW

6-1. The M5 Modular Crowd Control Munition (MCCM) is primarily a crowd control and area denial system that provides an alternative to lethal force. It is a nonlethal option used to control hostile crowds by temporarily incapacitating without causing life-threatening consequences. This gives commanders the option to apply nonlethal force as a first line of defense where appropriate. Appendix C describes a Practical Exercise for the M5 MCCM.

M5 MCCM SPECIFICATIONS

6-2. The M5 MCCM produces a flash-bang effect in addition to propelling 600 small plastic balls traveling at 155 meters-per-second. The M5 MCCM is effective against individuals wearing one layer of clothing at distances between five (5) to 15 meters. The probability of hitting four (4) out of five (5) E-type silhouette targets is 90 percent at five (5) meters and 80 percent at 15 meters. Basic operating temperature is from zero (0) degrees F to 120 degrees F, and storage temperature is -45 degrees F to 165 degrees F.

6-3. Engaging personnel closer than five (5) meters could result in serious injury or death. Do not engage personnel within five (5) meters.

6-4. The M5 MCCM comes packaged in a wooden crate. Inside the wooden crate is cardboard barrier packaging that contains six (6) bandoleers. Each bandoleer contains an instruction sheet sewn into the bandoleer, one (1) M5 MCCM, one (1) shock tube/blasting cap assembly, one (1) shock tube cutter, one (1) wire nut, and one (1) M81 igniter.

6-5. The M5 MCCM consists of the following components:

- Plastic olive drab green body and light green rear cover with raised diamond pattern to aid in the identification of the M5 as nonlethal munition when chemical protective gloves or in limited visibility.
- 600, 32-caliber plastic balls packed in a pliable gel arranged in two (2) layers of 300 balls each.
- 12 to 14 grams of a pentaerythritol tetranitrate sheet explosive cut into a pattern to provide maximum performance.

6-6. Additionally, thin foam spacers within the munition keep internal components snug within the munition module. The rear cover is glued to the rest of the munition and should not be removed or tampered with. The M5 NSN is 1377-01-464-2606. Figure 6-1 shows the M5 MCCM with components.

M5 MCCM COMPONENTS

- Six (6) bandoleers per box.
- Each bandoleer contains the following:
  - Instruction sheet (sewn into bandoleer).
  - One (1) MCCM.
  - One (1) shock tube / blasting cap assembly.
  - One (1) M81 Igniter, a cutter, and wire nut.

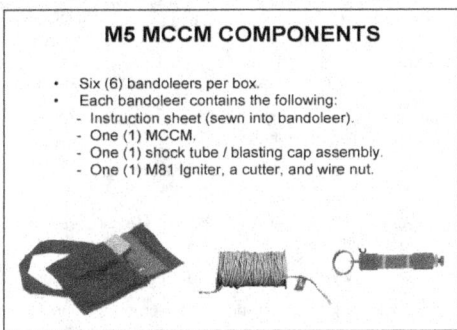

Figure 6-1. M5 MCCM with Components

Chapter 6

SHOCK TUBE/BLASTING CAP ASSEMBLY

6-7. The non-electric blasting cap assembly (shock tube) is used as the primary means of detonating the M5 MCCM. Because it is a non-electric component, this reduces the chance for accidental initiations due to static, transmitter, or electrical discharge. The energy created by the blasting cap is powerful enough to detonate standard military explosives. This component is pre-crimped to the blasting cap so there is no need for the user to crimp the components together. The opposite end of the shock tube is factory sealed with a heat sink to protect against moisture seepage. The shock tube is pre-spooled for ease of use.

6-8. The shock tube is a 110-foot-length, two-piece tubing with a polyethylene (soft plastic) outer wall and a surlyn (hard plastic) inner wall. The color of the live item is olive drab green. The interior of the shock tube is coated with a thin layer of an explosive material mixture of 90 percent high melting explosive and 10 percent aluminum powder. Once ignited, the explosive velocity travels through the tube at 6,500 feet-per-second (fps) to the blasting cap. The shock tube contains the explosive wave so the tube does not explode, get hot to the touch, melt, or change its shape. The only difference in the shock tube prior to and after igniting is at the blasting cap area or lack of cap. However, due to safety, do not hold the shock tube when firing it.

6-9. Prior to inserting into the M81 igniter, cut the shock tube approximately six (6) to eight (8) inches from the heat-crimped seal using the provided cutter. The cut should be a straight 90 degree cut approximately six (6) to eight (8) inches off the crimped/sealed end of the shock tube. Do not cut the tube on an angle as the M81 igniter may not ignite the shock tube.

> **WARNING: Do not strike, yank, or mishandle the blasting cap as this could cause premature detonation resulting in serious injury.**

Blasting Cap Demonstration

Note: Students should view a video clip of the blasting cap being detonated. Instructor shows a clip from a preliminary test of a non-electric blasting cap inside standard issue ammunition can. This is to show users how powerful the blasting cap is and to emphasize that the cap should be handled with care. The video clip is included in the MMTSP.

M81 IGNITER

6-10. The M81 igniter is used to ignite the shock tube component. Once the shock tube has been cut, loosen the end cap one half-turn. Remove the green solid plug, ensuring that the clear soft plug remains in place. Insert the cut end of the shock tube into the M81 igniter approximately one (1) inch. Next, tighten the collar of the M81 igniter. Remove the cotter pin safety and grasp the M81 igniter tightly. When ready to detonate the M5 MCCM, pull the ring on the M81 igniter sharply. It is recommended that the firer wear a leather-palmed glove on the hand holding the M81 igniter. Follow these steps to insert the M81 Igniter into the M5 MCCM.

1. Remove the anti-static bag containing M81 igniter, shock tube sealing nut, shock tube cutter, and supplemental instruction sheet from the bandoleer.

2. Cut anti-static bag one quarter-inch from the seal so bag can be resealed in the event of shock tube recovery.

3. Remove M81 supplemental instructional sheet, and shock tube cutter from anti-static bag. Leave shock tube sealing nut in bag in the event of shock tube recovery.

4. Loosen, but do not remove, M81 end cap by turning it counter-clockwise one half-turn.

5. Pull (do not twist) and remove the plastic weather-proofing plug from fuse cap holder.

Employ Modular Crowd Control Munition

6. Place in anti-static bag for possible reuse.
7. Place anti-static bag back in empty bandoleer pocket.
8. Using shock tube bag cutter, cut approximately six (6) to eight (8) inches (90-degree cut) off the crimped/sealed end of shock tube and discard.
9. Insert shock tube all the way into the hole from which the weather proofing plug was removed until it stops (approximately one (1) inch). Twist while pushing the shock tube in to ensure that it makes good contact with the M81 igniter primer.

Caution: End of shock tube must be seated fully into shock tube holder in order for M81 igniter to properly initiate shock tube.

10. Tighten the fuse holder in order for M81 igniter to properly initiate shock tube.

6-11. Hearing protection is required for all personnel within 300 feet (90 meters) of a detonating M5 MCCM. Figure 6-2 shows the M81 Igniter and highlights installing instructions.

WARNING: Lethal trauma may occur to individuals in front of munition if M5 MCCM is fired closer than 16 feet (5 meters).

M81 IGNITER

- Used to Ignite shock tube.
- Loosen collar 1/2 turn.
- Remove green plug and insert cut end of shock tube, (Make sure shock tube is inserted approximately one (1) inch), then tighten collar.
- Remove safety, pull ring quickly.
- Hold with leather work glove (optional).

Figure 6-2. M81 Igniter

M5 MCCM EMPLOYMENT CONSIDERATIONS

6-12. The M5 MCCM is a nonlethal variant of the Claymore munition and is designed to be a part of the NLCS. The M5 MCCM is a nonlethal munition used to incapacitate large groups of personnel with the flash bang and impact of rubber balls. It can be deployed by mounted or dismounted troops and provides a visual deterrent due to similarity in appearance to the M18A1 Claymore munition. The MCCM can be fired singularly or in a group and has a sweet spot of five (5) to 15 meters with 60-degree coverage. In the ground employment mode, the M5 MCCM will increase fixed-site and area security capabilities against potentially hostile forces without the sole reliance on the application of deadly force.

6-13. The M5 MCCM is intended to be a direct-fire, low-hazard munition that produces a nonlethal effect upon impact by incapacitating personnel through robust flash-bang and stinging plastic balls. The M5 MCCM uses 600 rubber balls (32-caliber) set in a two (2)-layer matrix of inert binder chemically similar to children's "play-dough." A sheet explosive 0.042 inches thick is used as the propellant. The sheet is sandwiched between the ball matrix and a polyethylene foam layer. The balls are launched in a fan-shaped distribution pattern with a maximum effective range of 15 meters.

6-14. The M5 MCCM looks similar to the M18A1 Claymore, except that the M5 MCCM is light green in color and the back cover has molded raised diamonds for tactile discrimination by troops wearing work gloves, and/or chemical protective gloves. The M5 MCCM is initiated using a blasting cap assembly supplied in the bandoleer. Figure 6-1 shows the M5 MCCM.

M5 MCCM SAFETY CONSIDERATIONS

6-15. Follow all local range regulations and ensure the proper wear of appropriate range uniform. Review safety issues found in FM 3-34.214, *Explosives and Demolitions*. Safety requirements include the following:

- Read packaged operator instructions prior to training with the M5 MCCM.
- Maintain the device in original packaging until ready to use.
- The instructor will maintain possession of the M81 igniter during training at unit level until ready to install the shock tube.
- Hearing protection is required.
- Report malfunctions or other unsafe conditions.
- Shock tube terminates at cap.
- Do not hold shock tube while firing.
- Do not disassemble the M5.
- Do not strike, drop or abuse blasting caps.
- Do not pull on the shock tube.
- Dispose of residue IAW appropriate regulations and local SOP.

6-16. Other safety considerations include the following:

- Know the proper aiming distance and angle.
- Know general ammunition safety procedures.
- Be familiar with the M5 MCCM.
- Maintain the M5 MCCM in its original packaging until ready to use.
- Report malfunctions or other unsafe conditions.
- Remember the shock tube terminates at the blasting cap.

- Do not hold shock tube while firing.
- Do not disassemble the M5 MCCM.
- Do not engage personnel with the M5 MCCM closer than five (5) meters.

PREVENTIVE MAINTENANCE CHECKS AND SERVICES

6-17. The M5 MCCM is a one-shot device; as such, operator and unit maintenance will be minimal. Used in the field, the device will not require maintenance beyond the current capabilities of assigned unit maintenance assets.

6-18. PMCS are limited to a visual inspection at the time of issue to ensure all components are present.

6-19. Shock-Tube Rejection Criteria. If the shock tube is not damaged to the core (indicated by presence of silver powder), the shock tube is serviceable. If the shock tube has minor damage, it can be reinforced with tape, if available.

6-20. Using a moist cloth, carefully remove any mud or debris from external surfaces of turned in, unused M5 MCCM or M81 igniters prior to repacking.

EMPLACING, AIMING, AND ARMING THE M5 MCCM

6-21. The MMTSP provides a video clip of the set up procedures described in this section.

6-22. Place the M5 MCCM in position and pace off five (5) meters. Place an object (stone, branch, etc.) at that point. This mark will serve as the minimum engagement line. Follow the steps described below for set up procedures:

- Remove the shipping plug priming adapter (either one will work).
- Insert the blasting cap from the Blasting Cap Assembly in the adapter. Slip the shock tube in the groove of the shipping plug.
- Thread the adapter back into cap well of the M5 MCCM. Ensure the blasting cap is well seated into the well. Failure to do so may result in a misfire.
- Check the aiming point of the M5 MCCM using the procedures described below.
- Carefully uncoil the shock tube assembly and place a rock or a sandbag on the shock tube to keep it stationary. Care should be taken not to anchor with anything heavy enough to crimp the shock tube.
- Continue uncoiling the shock tube back to the safe detonating distance. Safe detonating distances are 50 feet for a shielded firer and 100 feet for an unshielded firer.

NOTE: The MCCM should never be mounted to a vehicle.

AIMING THE M5 MCCM

6-23. Select an aiming point approximately 16 feet (five (5) meters) in front of the munition and approximately one (1) foot or boot/calf height. Follow these procedures for aiming the M5 MCCM:

- Unfold and press the legs of the M5 MCCM firmly into the ground (should be able to place fist between munition and ground).
- Position eye approximately six (6) to nine (9) inches to the rear of the sight.
- Aim the munition by aligning front and rear edges of sight with aim point.
- Align the front and rear sights to aim the M5 MCCM sight at the bottom of the object. Aim sector is between five (5) meters and 15 meters at a 45 degree angle in front of munition.

Chapter 6

6-24. Students view a video clip of the aiming procedures (provided in the MMTSP).

ARMING PROCEDURES

6-25. Using care not to damage the shock tube; secure the shock tube approximately 18 inches from the blasting cap side of the munition with a stone, gravel, or sand bag to prevent aiming munition from being misaligned if shock tube is disturbed.

6-26. Minimum safe distance between the munition and a protected or shielded operator firing position is 50 feet (15 meters), and 100 feet (30 meters) between the munition and an unprotected position.

6-27. Unravel remainder of shock tube until it reaches the protected or shielded operator firing position. Place empty tube spool into bandoleer pocket for reuse if munition is recovered.

> **WARNING:** Do not hold shock tube when firing. There is possibility of hot gaseous byproducts being released from the M81 igniter forward end opening (shock tube insertion point) when initiating the munition. During firing of the M81 igniter, keep hands clear of this area.

MISFIRE PROCEDURES

6-28. Misfire Procedures in a training environment:

- Once user pulls the M81 igniter and MCCM does not function, user will then re-cock the M81 and attempt to fire again (this can be done twice).

- User will remove M81 igniter and measure approximately three (3) feet (one (1) meter) down the shock tube.

- Cut the shock tube at that point with provided cutter (90-degree angle) and then cut an additional six (6) inches off. (Cut should be a straight 90-degree cut. Do not cut tube on an angle; M81 may not ignite the shock tube.)

- Conduct the blow test by blowing through the six (6)-inch long shock tube with the open end aimed toward the palm of user's hand. If user notices a small amount of silver powder, re-install a new M81 igniter, if available, and attempt to fire.

- If user does not notice any silver powder in the palm of his/her hand, repeat Steps 2 through 4 until the process produces silver powder (this can be done two (2) additional times or up to nine (9) ft of shock tube). Re-install a new M81 igniter and attempt to fire.

- If MCCM still does not function, a wait time of 30 minutes is mandatory (FM 3-34.214), after which the user can flag the M5 MCCM and notify Explosive Ordnance Disposal (EOD) for removal.

> **WARNING:** During blow test DO NOT ingest or inhale explosive powder in shock tube. After handling shock tube, wash hands before eating or drinking.

6-29. Misfire procedures in an operational environment.

- Once user pulls the M81 igniter and M5 MCCM does not function, re-cock the M81 igniter and attempt to fire again. The M81 igniter can be re-cocked twice. (The M81 igniter may require one quarter-turn to push ring all the way in.)
- If the user has an additional M81 igniter and time permits, user will remove that M81 and measure approximately nine (9) to 10 ft. down the shock tube (two (2) double arm spans). Cut the shock tube at that point with provided cutter (90-degree angle). Re-install a new M81 and attempt to fire. (Cut should be a straight 90-degree cut. Do not cut tube on an angle; M81 may not ignite the shock tube.)
- If additional M81 is not available flag the M5 MCCM and notify EOD for proper disposal.

HANDLING AND RECOVERY PROCEDURES

6-30. Treat all explosive components with care. Do not strike, drop, or abuse blasting caps. Detonated or burning explosives and plastics produce poisonous fumes; do not inhale fumes while igniting shock tube. Residue must be disposed of properly (according to local SOP). Remember that there is shock tube residue once expended, check with your ammunition supply point on what residue must be turned in. Do not breathe in when applying the blow test because you may swallow the explosive powder instead of blowing through the tube.

DISARM AND RECOVER

6-31. If mission dictates that the M5 MCCM must be disarmed and recovered, take the following steps:

- Remove safety pin (cotter pin) from anti-static bag in bandoleer.
- Align hole in M81 igniter with hole in pull ring shaft and insert safety pin through holes.
- Loosen, but do not remove, fuse holder by turning it counter-clockwise three (3) to four (4) full turns.
- Remove shock tube from M81 igniter.
- Install saved weatherproofing plug and tighten fuse holder cap by turning clockwise.
- Seal end of shock tube by folding over approximately two (2) inches from end. Secure fold by screwing shock tube sealing nut from anti-static bag in bandoleer over the fold.
- Place M81 igniter and shock tube cutter in anti-static barrier bag. Fold over and seal open end of barrier bag with tape, if available. Place barrier bag in bandoleer.
- Take the spool from the bandoleer and re-coil shock tube, moving toward the munition.
- Unscrew priming adapter on M5 MCCM to remove blasting cap assembly from detonator well.
- Remove shock tube from primer adapter by sliding it out through the slot in the adapter.
- Replace shipping plug end of combination shipping plug/primer adapter into detonator well on M5 MCCM.
- Recoil remaining shock tube (checking for damage) and secure blasting cap end of shock tube onto spool using tape, if available.
- Collapse and fold legs of M5 MCCM and replace in barrier bag. Fold over opening of barrier bag and seal with tape, if available.
- Return all items to the bandoleer.

PRACTICAL EXERCISE

6-32. The instructor, assisted by the AI, talks the students through the employment procedures. Follow the procedures described in Appendix C.

6-33. Inform the students that they will be required to employ the M5 MCCM. Conduct a safety briefing prior to the start of the practical exercise. Inform the students that they will have 20 minutes each to complete the practical exercise.

6-34. Allow 10 minutes at the end of the practical exercise to review, answer student questions, and correct student misunderstandings. Critique and assist students as necessary throughout the practical exercise.

Chapter 7

Employ 66 Millimeter Nonlethal Grenades

OVERVIEW

7-1. Nonlethal grenades can produce different effects that range from distraction to blunt trauma to delivery of smoke or anti-riot irritants. They are primarily designed to provide protection to friendly forces, control crowds, and denial by stopping, confusing, disorienting crowds without the use of deadly force.

7-2. The nonlethal grenades described in this chapter are launched from the M7 66mm grenade-launching system mounted on a variety of tactical vehicles. The M7 is an indirect fire support system that can deliver the M98 distraction grenade that creates a flash-bang effect, the L96 antiriot grenade, or the M99 blunt trauma grenade that creates a sting-ball effect.

66MM NONLETHAL GRENADE SPECIFICATIONS

7-3. The five nonlethal grenade types discussed in this TC include the following:

- L96A1, Anti-riot, Irritant (NSN 1330-01-459-4018).
- L97A1, Anti-riot, Practice (NSN 1330-01-459-4032).
- M90 Nonlethal Smoke.
- M98 Nonlethal, Distraction (NSN 1330-01-484-7773).
- M99 Nonlethal, Blunt Trauma (NSN 1330-01-484-7775).

THE L96 AND L97 GRENADES

7-4. The L96A1 is loaded with twenty-three canisters containing a CS compound (causes extreme irritation to eyes and mucous membrane). Length is 7.28 in. and weighs approximately 1.2 pounds. Electrical current ignites the grenade's propellant charge ejecting it out of the tube and down range. Grenade's rubber body ruptures in flight and disperses the 23 canisters. Each canister produces a cloud of CS smoke for approximately 10-12 seconds. The L97A1 is the same configuration as the L96A1 except it's the trainer and contains cinnamic acid (CA) as a simulant.

7-5. Safety warnings associated with the L96A1 and the L97A1 grenades include the following:

- CS and CA smoke can cause irritation to eyes, skin, and mucous membranes. When entering a CS or CA smoke cloud, wear appropriate face protection (M17 or M40 series mask).
- In dry grass area, burning grenades could cause fires, keep fire extinguisher handy.
- During training, single hearing protection must be worn within 16 meters (including vehicle occupants).
- Do not fire grenades within 150m (90°arc around firing vehicle) from personnel or equipment.
- Fire only at the 100m aiming bracket setting.

THE M90 GRENADE

7-6. The M90 grenade is loaded with three canisters providing a protective smoke screen for light vehicles during hostile situations. Each canister is filled with Terephthalic Acid (TA). Length is 9.9 in. and weighs approximately 2.86 pounds. Electrical current ignites the grenade's propellant charge ejecting 3 canisters out of tube and down range. The aluminum tube remains in the discharger after grenade has fired. Tube must be removed to reload.

7-7. Safety warnings for the M90 grenade include the following:

- Prolonged breathing of obscurant smoke can damage lungs. When entering smoke cloud, wear face protection.
- In dry grass area, bursting grenade payloads could cause fire, keep fire extinguisher handy.
- During training, single hearing protection must be worn within 0.5 meters (including vehicle occupants).
- Do not fire grenades within 75m (90°arc around firing vehicle) from personnel or equipment.
- Fire only at the 100m aiming bracket setting.

THE M98 AND M99 GRENADES

7-8. The M98 and M99 are loaded with three canisters and are used to disorient or disperse rioting crowds. Each M98 (audio/visual) canister delivers 170 dB and a quick burst of light. Each M99 (blunt trauma) canister has same characteristics as the M98, plus delivers 140 - .32 caliber plastic balls. Length is 9.9 in. and weighs approximately 1.5 pounds. Electrical current ignites the grenade's propellant charge ejecting the 3 canisters out of tube and down range. The aluminum tube remains in the discharger after grenade has fired. Tube must be removed to reload.

7-9. M98 and M99 safety warnings include the following:

- Bursting grenades may cause cornea or skin injuries at very close range. Personnel within 35 centimeters of operating launcher or 0.5m of a bursting canister should wear safety or ballistic eye protection, long sleeves, elevated shirt collar and Kevlar.
- Avoid looking directly at bursting grenades, temporary loss of vision is possible.
- In dry grass area, bursting grenade payloads could cause fire, keep fire extinguisher handy.
- During training, single hearing protection must be worn within 17m (vehicle occupants included).
- Do not fire grenades within 160m (90°arc around firing vehicle) from personnel or equipment.

NONLETHAL GRENADE LOADING AND UNLOADING PROCEDURES

7-10. Before loading 66mm grenades make sure vehicle is parked with the ARM/OFF switch in the OFF position (light will be off). Remove the discharge cover, save and visually check tubes for damage or debris.

7-11. Safety warnings include the following:

- Handle grenades with care.
- Do not drop or throw grenades.
- Do not use damaged grenades.
- Keep grenades away from electrical sparks and hot surfaces.
- Keep grenades sealed until loading.
- Never position any part of your body in front of the dischargers once loaded.
- Always ensure ARM/OFF switch is OFF before loading.

7-12. Remove and unpack the required number and type of grenades from their container. Keep grenades clean free from dirt and grease. Insert grenade into discharger tube (electrical contact first), push down gently until grenade's spring clip engages tip at the bottom of the tube. You will feel two distinct clicks. Rotate the grenade ¼ turn, ensuring an electrical contact. The M7 is now loaded.

7-13. To unload the M7 place the ARM/OFF switch to the OFF position. Remove the grenades from discharger tubes and return them to their original containers. Install the discharger cover on the tubes and secure grenades IAW unit SOP.

7-14. Safety warnings for unloading procedures includes the following:

- Electrical spark could cause grenades to launch prematurely.
- Ensure ARM/OFF switch is OFF before unloading.
- Keep grenades away from hot surfaces.
- Never position your body in front of the dischargers during loading and unloading procedures.

NONLETHAL GRENADE LAUNCHING PROCEDURES

7-15. Adjust the elevation lever on the aiming bracket for required launch distance in meters. Use the following settings:

- M98/99 50m, 75m, or 100m
- M90, L96A1 and L97A1 100m only

7-16. Place the ARM/OFF switch to the ARM position, indicator light will be lit. Press the FIRE button, this will launch all grenades loaded (1-4). Switch the ARM/OFF switch to the OFF position, light will go off.

7-17. When tactical situation permits, check to see if all grenades launched. If not, perform Hangfire, Misfire, or Dud procedures.

HANGFIRE AND MISFIRE PROCEDURES

7-18. A hangfire is a temporary failure or delay in the action of the propellant charge. If this occurs, wait 10 seconds and make two additional attempts to fire (within a 10 second interval). If the grenade fails to fire after waiting 5 minutes, treat as a misfire.

7-19. A misfire is the failure of either the L96A1 or L97A1 grenade to eject from the discharger OR the failure of the M90, M98 and M99 payload canisters to eject from the grenade tube.

> **WARNING:** Grenades could accidentally fire. to avoid possible death or injury, position your body away from the front of loaded discharger. If misfired grenade launches during unloading, personnel in the area could be killed or injured. Keep vehicle pointed down range until grenades are removed.

7-20. Ensure that the ARM/OFF switch is set to ARM and press the FIRE button. If the grenade does not fire, set the switch to the OFF position and make sure the grenade is firmly seated in discharger tube. Next place the ARM/OFF switch back to the ARM position and attempt to fire again. If the second attempt fails, return the switch to the OFF position, remove that grenade from the discharger and install grenade into another discharger tube.

7-21. When relocating the grenade, hold it away from body and point it downrange. Attempt to fire grenade from another discharger tube. If the grenade fires, notify the unit maintenance to repair or replace the defective tube. If the grenade still fails to fire, remove the grenade and place it 200 M (220 yds) away from personnel and equipment.

7-22. Notify EOD to dispose of grenade(s).

DUDS

7-23. An M90, M98 or M99 grenade dud is one that has fired its payload from the grenade tube but one or more canisters have failed to burn or explode. Wait 15 minutes and discard IAW unit SOP.

7-24. An L96A1 or L97A1 grenade dud is one that failed to dispense its payload once launched from the discharger tube. When training, wait 15 minutes and notify EOD to dispose of the dud.

Chapter 8

Employ Individual Riot Control Agent Disperser

OVERVIEW

8-1. The Individual Riot Control Agent Disperser (IRCAD) is an individual disperser for self-defense and keeping hostiles out of arms reach. It is designed to provide a safe and effective way to subdue a hostile subject without causing permanent injury.

8-2. The procedures described in this chapter apply to the M39 however, these procedures also apply to the M38 disperser and the M40 (Trainer).

8-3. Initial training for new personnel should be comprehensive and will include a level 1 contamination. Consider not certifying those who have any respiratory conditions such as asthma.

8-4. IRCAD training should include the following:

- Level 1 contamination (direct facial exposure).
- Unit policy and procedures.
- Employment practice using inert training device.
- First Aid and decontamination.

8-5. When planning or conducting riot control agent disperser training, ensure the trainers have the correct materials. Notice that the M39 IRCAD (Live Agent) is red with a white label and the letters "OC" printed on the front of the can, while the M40 IRCAD (Trainer) has a gray label with "Simulant" printed across the front of the can.

8-6. Sustainment training should be conducted on a "regular" basis as dictated by unit policy. Policy changes should be included in sustainment training. OC users should be made aware of any case law, ROE, or liability issues that may affect the use of OC. Discuss and critique incidents that resulted in the use of OC.

8-7. Sustainment training should include a Level 2 (rubbing a contaminated cloth over closed eyes) or Level 3 (walking through an OC mist) contamination followed up with fight through drills. Fight through scenarios teaches the users why and when to use OC, as opposed to other training that may only address how.

8-8. Fight through scenarios can be constructed with:

- Inert training units.
- Video presentations.
- Simulators such as Firearms Training Systems (FATS).

NOTE: Riot control agents will be employed only when authorized by the President and geographical combatant commander, subject to the effective ROE, and then only defensively, to protect U.S. personnel and installations.

DEFINITIONS

8-9. Trainers should be familiar with the following common terms and definitions associated with IRCAD:

- Oleoresin – A mixture of a resin and an essential oil occurring naturally in various plants.
- Capsicum – Any plant of the genus capsicum occurring in many varieties that range from mild to hot, having pungent seeds enclosed in a potted or bell-shaped pericap.

- Oleoresin Capsicum – Oil of capsicum.
- Pungency – The heat or intensity of the pepper.
- Capsaicinoids – A group of alkaloid compounds, naturally occurring within the fats, oils, and waxes of the pepper plant; the amount of these compounds determines the pungency of the pepper.
- Capsaicin – The most prevalent of the seven compounds found within the Capsaicinoids and considered the active ingredient in OC; these compounds can be measured in a laboratory using a method of analysis called High Pressure Liquid Chromatography.
- Scoville Heat Units – A scale used to define the perception of heat based upon the capsaicinoid content of the capsicum plant.
- Solvents – A liquid substance capable of dissolving or dispensing one or more other substances.
- Emulsifier – A substance that creates an emulsion, or a mixture of mutually insoluble liquids in which one is dispersed in droplets throughout the other; bonds two or more liquids together.
- Carrier – The ingredient or ingredients, other than the OC, which comprise the OC formulation.
- Propellant – The gas or liquid that pressurizes the canister and propels the carrier and agent to the target.

M39 IRCAD SPECIFICATIONS

8-10. The M39 is an IRCAD used for self-defense and for keeping rioters and/or hostile subjects out of arms reach. The M39 IRCAD contains enough OC or "Pepper Spray" for 15 one (1)-second bursts. It has an operational range of 10 to 30 feet and comes with a carry pouch. The M39 NSN is 1040-01-501-4380. The M40 Trainer NSN is 1040-01-501-4423.

8-11. The M39 IRCAD is divided into six different parts that include the following:

- Nozzle – Dispenses the product from the canister according to the prescribed pattern.
- Canister – Contains the product.
- Safety Cover – Plastic latch located on top of the actuator button.
- Actuator Button – Mechanism that activates the product.
- Valve Assembly – Connects the tube to the valve stem.
- Tube Delivery – Delivery system to the valve assembly.

Note: This is the nomenclature for the M39 IRCAD. Other products such as the M38 may have a slight difference in nomenclature depending on the manufacturer and the specifications required.

8-12. Figure 8-1 shows the M39 IRCAD.

M39, IRCAD DESCRIPTION

The M39 is an individual Riot Control Agent Disperser (IRCAD) for self-defense and keeping rioters and/or hostile subjects out of arms reach.

The M39 contains enough Oleoresin Capsicum (OC) or "Pepper Spray" for 15 one (1) second bursts. It has an operational range of 10 - 30 feet and comes with a carry pouch.

Figure 8-1. M39 Individual Riot Control Agent Disperser

SPRAY PATTERNS

8-13. Spray patterns are defined by how the OC is displaced when leaving the nozzle of the canister. There are three basic spray patterns used by all manufacturers: fog, stream, and foam. They will be employed according to the type of canister and the environment in which they will be used.

8-14. **Fog Pattern.** Fog (Cone/Mist) – Smallest particulate size. Handheld Fog/Cone spray patterns are dispersed in a wide formation (similar to a shotgun effect) making it easier to acquire the target.

- The spray is completely filled with microscopic droplets causing every area around the subject's eyes, nose, and face to be covered.
- Full cone patterns are affected more by wind conditions, and, due to the nozzle design, generally do not have as many spray bursts per canister as stream patterns.
- The minimum spraying distance is 36 inches with an effective range of three (3) to eight (8) feet.

8-15. **Stream Pattern.** Stream pattern is the larger particulate size. Concentrated stream allows a greater range in its delivery system. Use of the stream contains the contamination in a more concentrated area.

- The ballistic stream can be used to select an individual in a crowd with greater accuracy and reduce the likelihood of contaminating other subjects or Soldiers who may be in the area.
- This pattern hits the subject with a splash or splatter effect (dependent upon the distance), giving it an effective range of three (3) to 12 feet.
- The minimum spraying distance is 36 inches.

8-16. **Foam Pattern.** Foam pattern is the most concentrated particulate size.

- Jet foam patterns – A powerful fast-acting foaming surfactant that coats the face upon contact. This pattern hits with greater impact, has better surface adhesion, reduces cross contamination, and has an effective range of three (3) to five (5) feet.
- It is designed for climate-controlled environments such as courtrooms, hospitals, schools, and holding facilities. It is easier to see the application during low light conditions.
- The minimum spraying distance is 36 inches.
- Some throwback potential exists and may possibly be inhaled; the product may become slippery on smooth surfaces.

DELIVERY METHODS

8-17. The method of delivery is defined by how the OC is applied to the aggressor(s) depending upon the particular spray pattern. The three (3) most effective ways to deliver OC are:

- Up and Down – OC is dispersed by spraying in a sweeping motion from the eyes. This method of delivery is recommended with the fog spray pattern.
- Side to Side – OC is dispensed by spraying in a sweeping motion from ear to ear, concentrating on the eyes. This method of delivery is recommended with the stream spray pattern.
- Spiral Motion – OC is dispensed by spraying in a tight circular motion, concentrating on the facial area. This method of delivery is recommended with the foam spray pattern.

HYDRAULIC NEEDLE EFFECT

8-18. The hydraulic needle effect is an important factor to consider when employing OC. This is the consequence of the OC particulate penetrating the soft tissue of the eye. This is due to the correlation between the distance and the amount of pressure (size of the canister) in which it is delivered. Concerns have been raised about the possibility of soft tissue injury, prolonged irritation, or possibly infection.

8-19. Because of the possibility of the hydraulic needle effect, minimum safe distances have been established for each delivery system.

8-20. Instances of hydraulic needle effect are rare, but nevertheless should be taken into consideration.

8-21. Safety of the individual employing OC should never be compromised by delaying the use of OC in tactical situations for the concern of a hydraulic needle effect. Let the tactical situation determine the tactical response.

IRCAD EMPLOYMENT TECHNIQUES

8-22. This section describes drawing methods, grip methods, and stance when employing OC.

8-23. **Drawing.** There are three (3) basic ways of drawing the OC canister from the holster. Each method is acceptable; however, practice is recommended on each. Drawing techniques include the following three methods:

- Strong Side Draw – This is a draw where the canister is worn on the strong side of the user's body. The user will unfasten the top of the holster with the strong hand, remove the canister with the strong hand, and assume a ready position.
- Cross Draw – This is a draw where the canister is worn on the weak side of the user's body. The user will unfasten the top of the holster with the strong hand, remove the unit from the holster with the strong hand, and assume a ready position.
- Tactical/Assist Draw – This is a draw where the canister is worn on the strong or weak side of the body. The user will unfasten the holster with the weak hand, while simultaneously drawing the canister with the strong hand, and assume a ready position.

Note: As with a firearm or side handle baton, it is impractical to draw the OC canister with the weak hand; therefore, a weak side draw should not be used.

8-24. **Gripping.** Proper grip of the handheld OC canister is just as important as drawing the canister. Grip the canister using a "C" clamp. The fingers are extended firmly around the canister and snugly kept together with the thumb over the safety lid until ready to dispense. The index finger is under the nozzle guard. Actuation of the OC occurs by using the thumb or index finger, whichever feels most comfortable.

8-25. The benefits of using the thumb are often favored as it allows the user to apply direct pressure downward on the actuator for quick and smooth release of the OC.

- Advantage: If attacked while using the thumb to actuate the unit, the user could lift his thumb from the actuator and place his thumb back over the flip-top safety. This will increase the control the user has with his unit. It also allows the use of defensive or offensive hand techniques.
- Disadvantage: Some thumbs might not be able to fit in the actuator housing to allow the user to safely discharge the unit. Four (4) fingers can grasp stronger than three (3). Flexibility is the key!

8-26. Improper grip of the M39 canister could result in a very unfavorable outcome for the user. Proper grip is achieved when:

- The canister is held in the weak hand. The fingers are securely wrapped around the canister and held tightly against the strong side of the body.
- The strong hand grips the handle while the thumb is used to actuate the OC. If not held in this fashion, the aggressor may be able to grab the canister and detach it from the handle. If this should happen, it will cause the contents to completely engulf the user because of the pressurization and separation of the handle from the canister.

8-27. **Stance.** There are four basic stance positions.

- Two Hand Stance. The canister is held with the bottom of the canister over the user's forward foot. Do not fully extend the arm holding the canister. The weak hand is in a palm-down position on top of the strong wrist. This stance presents a dominant and authoritative appearance and alerts others that OC is being used.
- One Hand Stance. The canister is held with the bottom of the canister over the user's front foot. Do not fully extend the arm holding the canister. The weak hand is positioned either with the fingertips lower than the cheekbone and forward of the nozzle or behind the nozzle; in either position, do not extend the canister any further than three (3) to six (6) inches from the chest. This stance presents a dominant and authoritative appearance allowing for easy transitions between weapons and provides a clearing or checking hand.
- Two Hand Concealed Carry/Front Position. The user assumes a good, stable position with the strong leg back and the canister held in front of and close to the body. Both of the elbows remain above the user's duty belt, placing the free hand over the unit to conceal it from view. The thumb should be kept off the actuator and on the safety cap. This carry presents a professional appearance and a low profile approach for the user and will not alert other bystanders that the user is ready to employ OC. The designated finger needs to be above the flip-top safety to prevent accidental discharge.
- Low Profile Carry. The user assumes a good stable position with the strong leg back. The canister is held in the strong hand extended down to his or her side keeping the thumb on the safety cap and placing the knuckles of that hand to the center of the buttocks. Primarily used for approaching a subject from a concealed area. This carry presents a professional low profile approach for the user, which will not alert other bystanders that the user is ready to employ OC. The designated finger needs to be above the flip-top safety to prevent accidental discharge.

SECURING THE SUBJECT

8-28. After spraying subject, order the subject to a prone position. The contaminated subject will have the means to effectively resist. It is crucial to establish control by having them assume a prone position. This will give more control over the subject. Quickly place handcuffs or restraints on the subject, reassure them that they will be fine, and tell them what you will be doing to them.

8-29. Do not press down on the subject's back. Sitting, kneeling, or pressing on the subject's back will make it harder for them to breathe. Their airway will be restricted by the effects of OC. Pressing on their back could cause severe medical problems or death.

IRCAD SAFETY CONSIDERATIONS

8-30. During training have safety officers present for each individual contamination to assist in practical exercise, decontamination, and recovery phases. Instructors and safety officers must be cognizant of all students at all times.

8-31. Remove contact lenses prior to inert or "live" OC contamination. Cosmetics may either diminish or prolong the effects of OC and should be removed prior to contamination during training.

8-32. Do not contaminate students at a faster pace than the decontamination facility will accommodate. Medical assistance will be on scene and available during training where the contamination is executed. Consider placing medical personnel out of sight or at the water site so as not to heighten the anxiety of the students.

8-33. Training with inert units will help ensure the accuracy and effectiveness of individuals when employing OC sprays under a variety of conditions. Eye protection should be used when practicing with inert units directly on fellow students. The alcohol within the inert formulation may wash away the eyes' protective fluids.

8-34. The training area must have at least one adequate running water site in order to properly conduct the decontamination procedure. Using buckets of water in the decontamination process is acceptable only if it is continuously replenished.

8-35. The training area should be in an area away from onlookers to avoid unnecessary and inappropriate comments.

DECONTAMINATION, RECOVERY, AND FIRST AID

8-36. Once the subject has been restrained, begin the decontamination process. During transport reassure subject and monitor for distress, coherence, and respiration. Continually evaluate the subject for coherence and respiration. At any sign of distress, get immediate medical assistance.

DECONTAMINATION

8-37. General decontamination. Remove the subject from the contaminated area and establish a verbal rapport. Expose the subject to fresh air and have him/her face into the wind. Fans or air conditioning units may be used. Tell the subject to breathe in through the mouth, and out through the nose. Tell the subject to strobe the eyes (open and close rapidly). Do not allow the subject to rub his or her eyes.

8-38. If it is practical before transporting, apply immediate first aid/decontamination such as "COOL IT" or small water dispensers such as "FIRST RELIEF."

8-39. The most effective way to remove resin from the skin is by blotting the face with a wet paper towel followed by a dry one. Repeat this until the resin is removed.

RECOVERY

8-40. An individual will usually recover within one hour with vast improvements. The eyes should be able to open within 20 to 30 minutes.

8-41. Anyone not exhibiting significant improvement after one (1) hour should be closely monitored to ensure continued recovery.

FIRST AID

8-42. OC formulations, which exceed 0.60% Capsaicin, increase the potential for burns, particularly in fair-skinned persons (those who sunburn easily). Any person who exhibits sunburn-like redness more than one (1) hour after being decontaminated or who show any evidence of blistering (second degree burns) after being sprayed should receive medical treatment for burns. Avoid salves and ointments until affected area has been completely decontaminated.

8-43. Once a subject has been restrained after being sprayed, the user should conduct a Primary Medical survey: airway, breathing, and circulation.

- Open the airway.
- Check for signs of obstruction in the mouth.
- Check for signs of responsiveness.

8-44. No person who has been contaminated by OC or any other chemical agent should be left unsupervised for at least two hours after contamination. Medical personnel should immediately evaluate any person who admits to being under the influence of any drugs or alcohol. Medical personnel should evaluate any person who admits a history of heart problems, lung problems, diabetes, high blood pressure, or any other potentially serious medical condition.

CHARACTERISTICS OF OC

8-45. Because the OC formulation is heavier than air, the vapor rate of OC is very low and minimizes the possibility of transfer or cross contamination. Vaporization is when a substance changes from a liquid to a gas state and should not be confused with very small droplets or particulate that may remain airborne, such as a fogger.

8-46. These airborne particulates may move across rooms or through ventilation systems and are most prevalent in the following:

- Fog delivery systems.
- Spray nozzles that use conical spray patterns.

8-47. Several environmental factors should be considered when using OC in a tactical situation. These factors include the following:

- Wind and rain.
- Fans and ventilation.
- Heat and humidity.

FLAMMABILITY

8-48. Flammability and carcinogenic properties. Depending on whether a product is oil- or water-based, there will be a specific requirement for solvents and emulsifiers to ensure even suspension of the Capsaicin. These ingredients make up the majority of the formulation and should be closely evaluated for their safety.

8-49. Guardian Protective Services' OC products are nonflammable and non-carcinogenic. Although propylene glycol (emulsifier) and ethanol (solvent) are used in Guardian products, they are not used in sufficient quantities for the formulation to be carcinogenic or combustible if it comes into contact with a flame or a source of heat. Guardian OC products meet the non-flammability and non-carcinogenic requirement set forth by Army policy.

AWARENESS

8-50. OC should be used early with the element of surprise and prior to escalation of physical contact. Communicate with fellow Soldiers when spraying a subject who is in the proximity of or in physical contact with another Soldier. Use code words such as "spray" or "OC." Avoid using words or phrases that incite rage or anger like "nuke him."

TARGET AREA SPRAY VOLUME

8-51. The primary target when employing OC is the facial area, guaranteeing coverage of the eye zone (eyes, forehead, and brow). The secondary target is the nose and mouth. Discharge into facial area using as much as required.

8-52. If the open eye is contaminated, a half- to one-second burst should be adequate to achieve the desired effects. However, expecting an individual to accurately employ a projector for two bursts of a half second to one second during a confrontation may be unrealistic.

8-53. In situations where the subject is hit around the eyes (i.e., forehead or cheek), an ample amount of formulation should be employed to ensure that enough fluids are present to carry the OC particulate into the eyes. For multiple opponents, use as much as required to control the situation based upon the threat.

LEVELS OF CONTAMINATION

8-54. When handling or using any type of chemicals, there are three (3) levels of contamination. Each level will affect an individual differently. The three (3) levels of contamination are as follows:

- Level 1: This is defined as direct physical contact with OC.
- Level 2: This is defined as an indirect or secondary contact with OC. A level 2 contamination is the result of attempting to control or physically touch another person or item that has had a level 1 contamination. Moving in to control an aggressor who has just received a level 1 contamination may result in a level 2 contamination to the individual employing the OC.
- Level 3: This is defined as an area contamination with OC, such as after using it in an aerosol form. Usually, level 3 contamination will occur when entering a contaminated zone or area.

PHYSICAL AND PSYCHOLOGICAL EFFECTS OF OC

8-55. Knowing the physical and psychological effects will allow the control force member to know when it is best to employ the OC against an individual. Failing to take the effects of OC into consideration when attempting to control a crowd could end up being counterproductive to the goal of crowd dispersal.

PHYSICAL REACTIONS TO OC

8-56. Common physical reactions to OC include the following:

- Involuntary closure of the eyes, resulting in temporary visual impairment.
- Eyes will close when Capsaicin contacts the nerve endings.
- Eyes will remain closed due to the drying of the natural protective fluid of the eyes.
- Involuntary extension of the hands to the facial area.
- A burning sensation and inflammation of the eyes, mucous membranes, and a burning sensation to contaminated skin and tissues.
- The secretion of excessive mucous from the nose.
- Shortness of breath:
  - Capsaicin's inflammatory properties are a result of dilating blood vessels in the affected area. This action increases blood flow to the area, resulting in minimal swelling.
  - When this occurs within the nasal passages, the physiological effects trigger a psychological response, "I can't breathe."
  - The perceived inability to breathe can trigger a panic response, which manifests itself into hyperventilation.
- Approximately .005 of the general population may have an allergic reaction to various types of peppers.

- While most allergic reactions are not life threatening, it is necessary to provide medical treatment to any person believed to be having an allergic reaction.
- Any person who has been contaminated by an OC product who complains of itching, hives, difficulty swallowing, or facial swelling should be evaluated by medical personnel without delay.

PSYCHOLOGICAL EFFECTS OF OC

8-57. Anxiety is the fear of the unknown. It is normal for an individual to experience increased anxiety when faced with the unknown, such as being contaminated by OC for the first time. Some individuals may have an anxiety attack, causing them to change their breathing rhythms. Anyone who has never been contaminated with OC may display anxiety prior to contamination based on hearsay or rumors of its effects.

8-58. Fear is the confirmation of the unknown. It is normal for an individual to experience fear before, during, and after any physical confrontation. Individuals who have never been contaminated with OC may have their own fears and a premonition of what it does. Panic is the reality of one's fear. Some individuals may panic and flee without thought for obstructions or trip hazards.

Caution: Personnel who have never been contaminated by OC may panic if they are accidentally contaminated during the employment of OC. Train for the worst-case scenario.

EFFECTIVENESS

8-59. The failure rate of OC is difficult to quantify; however, it does exist. OC has a varied reaction time, ranging from one (1) to five (5) seconds. The mental state of an individual may be a significant factor to consider. Some people have a very high threshold for pain, especially subjects who are emotionally disturbed or prone to substance and/or alcohol abuse.

8-60. Mindset may influence effectiveness. Goal-oriented and mentally focused individuals may still accomplish their goal even though they cannot see and are experiencing significant discomfort. Many failures are operator errors due to lack of training or improper use of an OC product. OC is a viable force option when used by properly trained individuals and in conjunction with other force options.

ENVIRONMENTAL AND STORAGE CONSIDERATIONS

8-61. OC is biodegradable and does not require special equipment process for decontamination. With normal ventilation or by using high-speed fans, buildings, rooms, and vehicles can be decontaminated in approximately one hour. Ingredients may be washed down in drains. Blot exposed surfaces, clean with damp rag and non-oil-based soap. Clothes may be laundered as normal with other clothing.

STORAGE

8-62. Aerosol canisters may burst if exposed to temperatures above 120 degrees Fahrenheit (F). Prolonged exposure to temperatures below freezing (32 degrees F) may result in slower discharge. Canisters should be stored off the ground at comfortable room temperatures. Place stored aerosol canisters upside down to keep seals moist. Rotate the canisters upside down to right side up every 30 days. This prevents the seals from cracking and the OC canister from leaking.

8-63. Canisters should be physically inspected daily for damage. If carried on the duty belt unused for 30 days or more, discharge the OC in a designated area for a ¼ second burst. (This clears the nozzle of any debris or condensation.)

8-64. When stored in the armory, store the canister in the carrier with the nozzle towards the belt flap to keep any obstructions from getting into the nozzle.

CANISTER REPLACEMENT

8-65. OC canisters must be replaced at the end of the shelf life. Normally, the shelf life expires four (4) years after the date of manufacture. This is due to the potential loss of propellant, resulting in a slower discharge.

This page intentionally left blank.

Chapter 9

Employ Electro-Muscular Disruption Device

OVERVIEW

9-1. The primary electro-muscular disruption device (EMD) currently used by U.S. Army personnel is the X26E TASER. This device causes electro-muscular disruption as a safe and effective way to incapacitate personnel at distances up to 35 feet. A pronged dart system strikes anywhere on the body and delivers 50,000 volts, causing instantaneous incapacitation to the targeted individual. Appendix D describes the TASER Performance Evaluation.

ELECTRO-MUSCULAR DISRUPTION SPECIFICATIONS

9-2. X26E is a device used to fire two (2) barbed projectiles (probes) into a target with the purpose of delivering an electrical shock to subdue the target without harm to the target or the operator. The EMD weapons stun and override the sensory and motor nervous systems causing uncontrollable contractions of the muscle tissue.

9-3. The X26E is operated much like a standard issue sidearm complete with a laser sight. When the trigger is pulled, the two barbs are propelled by compressed nitrogen. The barbs are connected to the device by insulated wires and are effective within two (2) inches of the subject's body (the electrical charge can penetrate up to two (2) inches of cumulative clothing).

9-4. When the circuit is completed (both barbs must hit the target), a five (5)-second cycle of electrical pulses is initiated and sent through the wires into the subject's body. The cycle can be stopped by positioning the safety down to SAFE or can be reapplied by pulling the trigger a second time.

9-5. Figure 9-1 illustrates the various parts and functions of the X26E TASER.

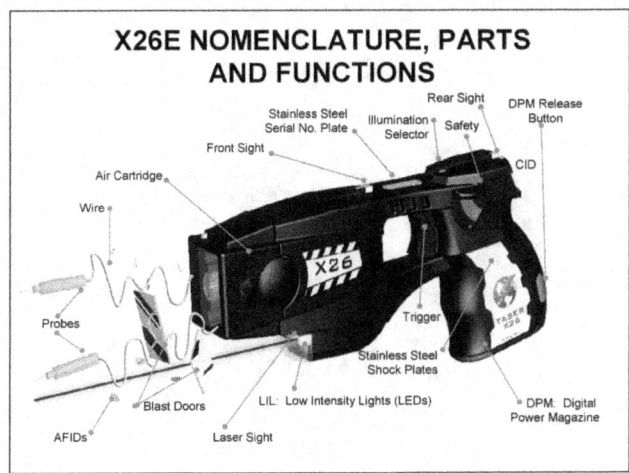

Figure 9-1. X26E Parts and Functions

9-6. X26E Nomenclature and Parts:
- Rear sights.
- Front sights.
- Safety—Ambidextrous.
  - Switch down: SAFE.
  - Switch up: ARMED; Activates Central Information Display (CID) and selected illumination.
- Grip.
- Trigger.
- Digital power magazine (DPM).
- DPM release button.
- Stainless steel shock plates.
- Low intensity illuminators.
- Laser sight.
- Contacts.
- Stainless steel serial number plate.
- Central Information Display.

X26E HOLSTER

9-7. The X26E ships with a holster that is ambidextrous and can be reconfigured for left-hand carry in minutes with an included hex key. Figure 9-2 illustrates the X26E with holster.

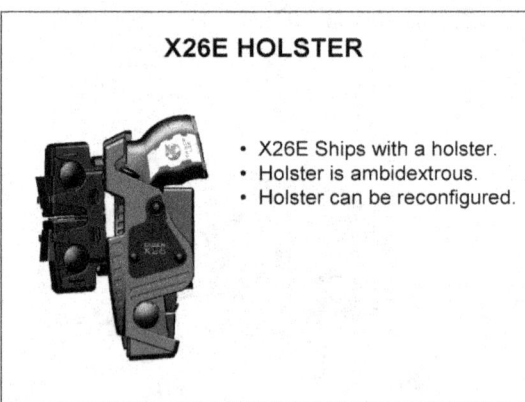

Figure 9-2. X26E with Holster

X26E ELECTRICAL SIGNALS

9-8. The X26E sends electrical signals similar to those used by the brain to communicate with the body. The signals overpower the normal electrical signals within the body's nerve fibers. The targeted subject instantly loses control of his body, cannot perform coordinated action, and usually falls to the ground.

9-9. The X26E sends out short-duration, high-voltage electrical waves that overpower the normal electrical signals within the nerve fibers. These waves create extra "noise" within the nervous system much like static on the "phone lines" in the human body. Stun weapons jam the central nervous system with electrical noise. This only affects the sensory nervous system creating pain compliance. The X26E can function as a stun weapon when used in the "drive stun" mode—the contacts placed directly in contact with the target. The X26E does not rely on pain to achieve compliance, but instead, overwhelms the central nervous system to achieve incapacitation.

X26E ELECTRICAL CURRENT

9-10. An electrical current of less than 0.02 amps can produce sensations ranging from tingling to sharp pain. A more serious effect occurs if the current causes muscles to contract. A person touching a live wire with their outstretched hand may literally not be able to let go of the wire due to the current's effect on the muscles. Currents from 0.03 to 0.07 amps will begin to impair the ability of the person to breathe. The X26E deliveries less than 0.004 amps.

9-11. Amps are the total electrons-per-second. The X26E output is less than 0.004 amps. Volts are the "pressure" pushing electrons. The X26E output is 50,000 volts. Typically, we experience 35,000 to 100,000 volts in our daily lives with "static electricity" when a spark jumps to us from door knobs, each other, etc. The danger is the amps, not the volts.

X26E ELECTRIC ENERGY

9-12. One (1) probe can arc through two (2) cumulative inches of clothing or one (1) inch of clothing per probe, including some bullet resistant materials. It is the high voltage that makes penetration possible.

9-13. Electricity (X26E energy) must be able to flow between the probes or the electrodes. Electricity (X26E energy) follows the path of least resistance between the probes. The greater the spread between the probes, the greater the effectiveness. The X26E energy will not pass to others in contact with the subject unless contact is made directly with probes or wires.

COMMON EFFECTS OF X26E EMD

9-14. Common effects of the X26E EMD include the following:

- Subject can fall immediately to the ground.
- Subject may yell or scream.
- Subject may experience involuntary muscle contractions.
- Subject may freeze in place with legs locked.
- Subject may feel dazed for several seconds/minutes.
- Subject may experience potential vertigo.
- Subject may experience temporary tingling sensation.
- Subject may experience critical stress amnesia (may not remember any pain).

X26E RISKS AND SIDE EFFECTS

9-15. Probable X26E EMD risks and side effects include the following:

- Can cause eye injury if shot too high.
- Can ignite flammable liquids or gases.
- Might cause slight signature marks that resemble minor surface burns (appear red or may blister).
- Can cause strong muscle contractions.
  - The exertion experienced is similar to athletic activities, such as weight lifting or wrestling and may result in similar type injuries such as muscle or tendon strain or tear, joint injuries, back injuries, stress fractures, or other secondary injuries resulting from strong muscle contraction.
- Muscle contractions may pose additional risk to certain persons, such as pregnant women.
- Can cause secondary injuries from person falling. Fall injuries, particularly from elevated heights, can pose risk of significant injury or death.
- Can cause pain and associated stress.

What the X26E Does Not Do

9-16. The following is a list of effects not common to X26E EMD:

- Does not damage nerve tissue.
- Does not cause "electrocution" in a wet environment.
  - Exposure to water will not cause electrocution or increase the energy to the subject.
- Generally does not cause urination or defecation.

DIGITAL PULSE CONTROLLER

9-17. The Digital Pulse Controller (DPC) is an internal circuit including the microprocessor of the X26E and various support hardware.

9-18. When the X26E is fired, the DPC measures the time between each energy pulse discharged from the weapon. The DPC then regulates the power throughput of the pulse generator to maintain a constant pulse rate. All pulse rates are approximate and may vary slightly.

9-19. The DPC automatically delivers a five (5)-second burst for each pull of the trigger. The DPC uses a variable pulse rate and lower DPMs in version 14. During the first two (2) seconds of each burst, the DPC runs at 19 pulses-per-second for maximum takedown power. After two (2) seconds, it slows slightly to 15 pulses-per-second for the remaining three (3) seconds in each burst. This lower pulse rate extends battery life by 25%. If the operator continues to hold down the trigger through the full five (5) seconds, the pulse rate will stay on at 15 pulses-per-second until the operator releases the trigger. In version 15 and higher, the pulse rate is 19 pulses-per-second for the full five (5) seconds.

9-20. The user has the capability to download X26E usage data using the data port download. The information that is downloaded is: Shot Sequence #, GMT Time, Local Time, Duration of the Shot in Seconds, Temperature of the Unit (deg C), and % of Battery Life Left. See Download Usage Data section at the end of this chapter.

CENTRAL INFORMATION DISPLAY

9-21. The CID displays the DPM power level, burst time countdown, warranty/general systems status, re-arm required status, and illumination setting.

9-22. The CID is a two-digit display on the back of the X26E. The CID communicates the following information:

- Zero (0)–99% DPM power level (energy cell indicator).

- When the safety is positioned upward to arm the weapon, the CID will display the percentage of DPM power remaining. This indication will last for five (5) seconds.

- After five (5) seconds, the CID will display two (2) dots to indicate the weapon remains armed. After 20 minutes, the system will shut down, and the CID screen will go blank. The unit WILL NOT FIRE. This condition requires re-arming the X26E by moving the safety from FIRE to SAFE and then back to FIRE. This is a battery saving strategy built into the device's programming.

- Burst Time Countdown. When the X26E is triggered, it delivers a five (5)-second energy burst. The CID displays a countdown from five (5) to zero (0) indicating how many seconds remain in the current burst. The burst can be stopped at any time by positioning the safety down to SAFE.

ILLUMINATION SETTINGS

9-23. The X26E is equipped with a laser sight in addition to two (2) low-intensity lights. When the Illumination Selector Switch is used (with safety on), the CID will display one of four codes to indicate which illuminators will be activated upon arming of the X26E. To cycle through the illumination options:

- Place X26E on SAFE and ensure it is unloaded.
- Press illumination selector and hold for one (1) second.
- Press and release to toggle modes displayed in CID.
    1. LO: Laser Only will illuminate.
    2. OF: Only Flashlight will illuminate.
    3. LF: Laser and Flashlight both illuminate.
    4. OO (Off/Off); Neither laser nor light will illuminate, and the CID goes dim.

9-24. System diagnostics can also be read through the CID. When a DPM is loaded with a warranty expiration date, current date and time, current Celsius internal temperature, and software revision level appear and flash in sequence.

DIGITAL POWER MAGAZINE

9-25. The DPM has a 10-year shelf life. Power comes from the lithium energy cells inside and provides up to 195 five (5)-second cycles (depending on the software version). A memory chip tracks the percent of energy left (% life). The DPM stores the percent of life remaining digitally and can be removed and used in other X26Es while still retaining its remaining power. The DPM must be left in the weapon at all times to maintain the system clock.

CUT AWAY OF THE X26E CARTRIDGE

9-26. Cartridge characteristics include the following:

- Solid yellow blast doors; live cartridge; 15-foot wires with regular probes.
- Striped blast doors; live cartridge; 21-foot wires with regular probes.
- Yellow cartridge with striped blast doors; live cartridge; 21-foot wires with longer, heavier probes (XP probes).
- Blue cartridge with striped blast doors; training simulation cartridge; 21-foot non-conductive lines with short probes (LS probes).

Chapter 9

9-27. Figure 9-3 illustrates a cut-away view of the X26E cartridge. AFID (seen in the illustration) refers to the anti-felon identification tag. The AFID system is explained later in this chapter.

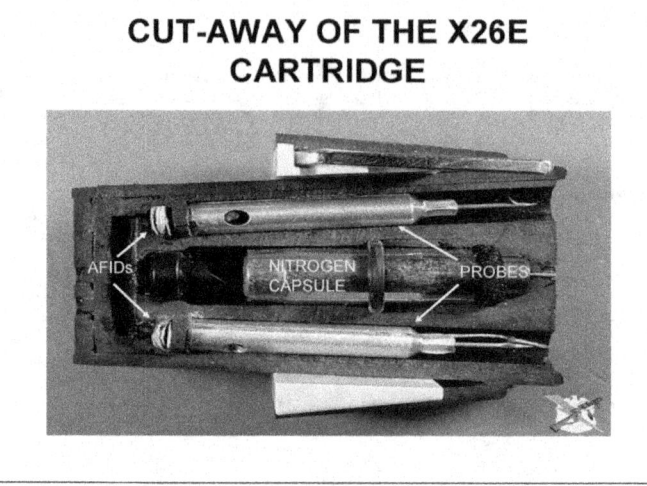

Figure 9-3. Cut-Away View of the X26E Cartridge

X26E CARTRIDGE CHANGES

9-28. This section describes the characteristics of various X26E cartridge changes.

- 21 feet (6.4 meters), striped door, live cartridge, regular probe, before August 2004.
- 21 feet (6.4 meters), silver door, live cartridge, regular probe, after August 2004.
- XP 21 feet (6.4 meters), yellow cartridge, live cartridge, XP probe, longer, heavier, before October 2004.
- Hybrid XP 25 feet (7.6 meters), green door, live cartridge, XP probe, heavier, after October 2004.
- LS 21 feet (6.4 meters), blue cartridge, striped doors, live simulation, short probe, non-conductive wire, stun portion works.
- Blue cartridge, blue door, live simulation, short probe, non-conductive wire, stun portion works.

XP35 Special Duty Cartridge

9-29. The 35-foot cartridge is a "special duty" cartridge, known as XP35. It requires extra training and practice firing to use this cartridge. Unlike the other TASER cartridges, this cartridge does not offer reversible loading on the weapon. The probes are set at a different angle than all other TASER cartridges. There is a top probe and a bottom probe. The top dart bore points up one (1) degree in relation to the device. Bottom dart bore is positioned four (4) degrees down from top dart bore. This was done for purpose of decreasing drop and spread amount in long distance shots. If the cartridge is not loaded properly, the point of aim/point of impact of the probes will be greatly affected. At distances less than 25 feet, the top probe travels above the laser.

9-30. This chamber is used for the XP35 cartridge in conjunction with a long-range cassette case that is not reversible. It has raised orange arrows printed on the side of the cartridge that is inserted toward the top of the device.

Probe Types

9-31. This section describes the various probe types that are currently available for the X26E TASER.

- LS probe:
    - Weight: 1.6 grams.
    - Length: 0.20 inches (0.50 centimeters).
    - Velocity at launch: 166 feet-per-second.
    - Velocity at 13 feet: 98 feet-per-second.
- Regular probe:
    - Weight: 1.6 grams.
    - Length: 0.35 inches (0.89 centimeters).
    - Velocity at launch: 166 feet-per-second.
    - Velocity at 13 feet: 98 feet-per-second.
- Hybrid XP probe is made of brass and aluminum, which is much denser than the aluminum of the other probes:
    - Weight 4.1 grams.
    - Length: 0.52 inches (1.32 centimeters).
    - Velocity at launch: 100 feet-per-second.
    - Velocity at 13 feet: 76 feet-per-second.
- The decay in velocity is primarily due to the wires uncoiling and creating drag on the projectiles.

9-32. Figure 9-4 shows the different probe types:

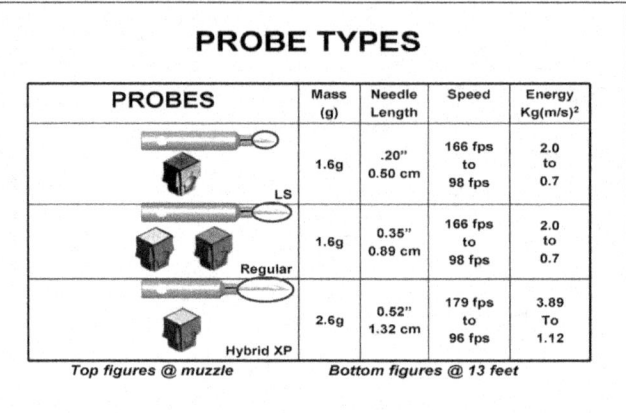

Figure 9-4. X26E Probe Types

PROPULSION SYSTEM

9-33. Characteristics of the propulsion system for 15- and 21-feet cartridges:

- 1,800 pounds per square inch (PSI) nonflammable nitrogen capsule.
- Two (2) probes fired at 160+ feet-per-second.
- Maximum range is 15 feet or 21 feet.

9-34. Characteristics of the Propulsion System for 25-feet Cartridge:

- 2,200 PSI nonflammable nitrogen capsule.
- Two (2) probes fired at 179+ feet-per-second.
- Maximum range is 25 feet.
- The pressure of the compressed nitrogen was increased to 2,200 PSI in the Hybrid XP cartridge only. This increases probe speed leaving the cartridge at over 179 feet-per-second and enables the probes to remain accurate out to maximum range.

X26E Wires

9-35. The wires are thin with insulated coating that can break easily if stepped on or pulled. Contact with the wires or the probe(s) during discharge can result in electrical shock. It is, therefore, critical that the X26E operator advise others in supporting roles to avoid wires during restraint for wire integrity. The effect of contact with a wire or probe while taking a subject into custody is relatively minor and will not cause EMD to the operator or those in support. Usually, operators will instinctively pull their hand away. It is recommended that operators control the subject in a different area of the body, away from the probes and the wires.

ANTI-FELON IDENTIFICATION TAGS

9-36. Anti-Felon Identification (AFID). Every time a cartridge is fired, it disperses 20–30 identification tags called anti-felon identification. These tags are printed with the serial number of the cartridge and can be used to determine who fired the cartridge. Operators should be aware that this system is an additional method of accountability to trace operators who are not following policy and are using the X26E inappropriately.

LOADING AND UNLOADING PROCEDURES

> **WARNING:** As soon as the X26E is loaded with a cartridge, it is a "live" weapon. Treat it as a loaded firearm. Do not allow or tolerate operators to be complacent with regards to safety just because it is not a conventional firearm. Do not point at people unless intending to fire. During practice, ensure the cartridges used are expended.

SAFETY

9-37. Standard weapons handling safety procedures apply. Always place safety switch in the down (SAFE) position; keep fingers clear from blast doors; do not place hand in front of the weapon while loading or unloading; keep the X26E pointed in a safe direction at all times; keep fingers off the trigger unless firing.

9-38. Never attempt to open a cartridge prior to firing. Tampering with a live air cartridge could cause it to fire or malfunction.

LOADING PROCEDURES

9-39. The cartridges are very easy to load and unload. To load, simply take a cartridge and snap it into the front of the unit. The air cartridge has been specifically designed so there is no "up" or "down" position enabling you to quickly reload it in a stressful situation, without worrying about putting it in upside down.

Improper Handling Technique

9-40. Figure 9-5 shows images of what happens when an X26E operator places his/her hand in front of the cartridge and the trigger is pulled.

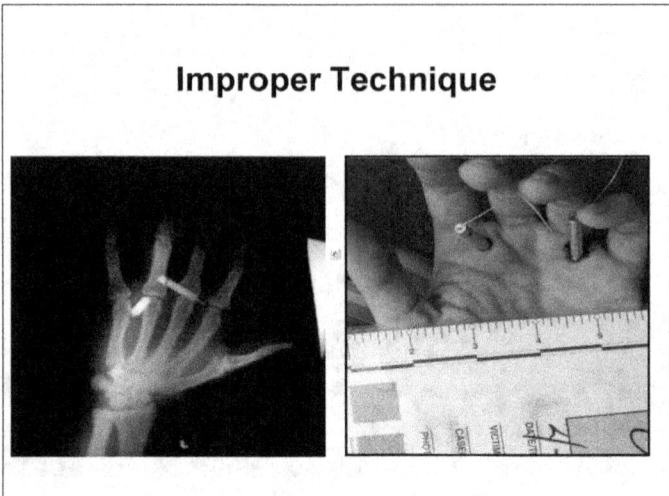

Figure 9-5. Results of Improper Handling Technique

UNLOADING PROCEDURES

9-41. To unload the cartridge, first make sure the safety is down in the SAFE, locked position. If you touch the cartridge while the unit is firing, you will be shocked by the stun electrodes on the front (the spark can jump from the contacts to your hand or fingers if they are too close). Once the safety is down, simultaneously squeeze both round release buttons on both sides of the cartridge and pull it out. You can now reload with another cartridge. The probes, the compressed air, and the wires are all contained in the cartridge; change only the air cartridge to be ready to fire again.

> **Caution:** With or without expended cartridges, the X26E will fire its electrical energy if the safety is off and the trigger is pulled. The X26E will cycle, spark, and expose you to unintended EMD and/or stun effect if you do not handle it properly. Keep hands, fingers, and any other parts of your body away from the contacts/cartridge when the safety is off or when the X26E is discharging.

PREVENTIVE MAINTENANCE, CHECKS, AND SERVICES

9-42. PMCS procedures must be followed to ensure safety and proper function of the X26E TASER.

9-43. If an X26E has been dropped or otherwise damaged, or if an X26E is exposed to significant moisture, do not move the safety switch to the up (ARMED) position until after conducting the following check. Failure to perform this check may result in an unintentional discharge when the X26E safety switch is placed in the up (ARMED) position.

9-44. Some X26Es exposed to extreme moisture have discharged with the safety switch still in the down (SAFE) position due to short circuiting of the electronic components.

9-45. Dropped or wet X26E PMCS procedures include the following:

- Safety switch down (SAFE).
- Remove the cartridge.
- Dry the device thoroughly.
- Safety switch up (ARMED).
- If device discharges without pulling the trigger, remove DPM and return the X26E to supply.
- Functions check—spark test for three (3) full five (5)-second cycles.
  - If the device does not function properly, return to supply.
  - If spark test/functions check is normal, return to service.

9-46. Inform operators to keep the X26E from getting excessively wet. The X26E is an electrical system. Hence, it is complex and can experience different failure modes under extreme conditions. For example, an EMD device that was exposed to an intense "salt fog" test spontaneously fired a cartridge. The salt fog condensed on the trigger switch, which eventually shorted the switch and fired the weapon. At least one such failure was also reported from the field. While the manufacturer has taken corrective actions to mitigate this risk, it cannot be completely eliminated.

9-47. X26E operators have experienced spontaneous discharges with and without cartridges loaded. A common denominator to those spontaneous discharges has been the presence of moisture (rain) or a static electricity source (spinning aircraft rotors). The X26E must be carried in a holster for ready access and protection from the elements, while being capable of completely containing a cartridge discharge.

9-48. If an X26E has been dropped, damaged, or exposed to significant moisture, take the following steps:

- Remove the cartridge immediately.
- If exposed to moisture, dry the weapon thoroughly (at least 24 hours) before proceeding. After 24 hours, ensure that all components are completely dry.
- Remove DPM from the weapon.
- Wipe off all exposed surfaces.
- If there is any visible moisture inside the DPM well of the X26E, contact TASER International for a Return Materials Authorization and send your X26E in for inspection. If no moisture is visible, re-insert the DPM and complete the remaining steps.
- Wait one minute before proceeding to the next step.
- Place the safety switch in the up (ARMED) position.
- If the X26E discharges without pulling the trigger, place the safety switch in the down (SAFE) position, remove the DPM, and contact TASER International for a Return Materials Authorization. DO NOT reinstall the DPM or air cartridge or attempt to use the device.
- If the X26E does not discharge without pulling the trigger, conduct a spark test for three (3) full five (5)-second cycles to ensure a rapid pulse rate and that the discharge stops after five seconds.
- If the X26E does not operate properly, place the safety switch in the down (SAFE) position, remove the DPM, and turn it in to supply.
- If the X26E does function properly, place the safety switch in the down (SAFE) position and return to normal use.

9-49. In addition to the above test, a thorough inspection of a X26E after being dropped is required. Check the entire X26E for cracks, broken CID, cracked laser/flashlight lens, etc. If any damage is found, contact TASER International for a return authorization. DO NOT use an X26E that shows obvious signs of damage.

9-50. Avoid dropping the X26E. It is a sensitive, electronic device and should be provided the same care and consideration as a laptop computer or cell phone.

- Check DPM power level regularly.
- Always store the X26E with DPM inserted.
- Check expiration of air cartridges.
- Secure in protective holster when not in use.
- Do not store/carry in pockets without a holster.

9-51. Occasionally wipe out the cartridge firing bay with a dry cloth. Multiple cartridge firings create carbon buildup (particularly after training courses).

DPM REPLACEMENT

9-52. Replace DPM when battery life is less than 20 percent. Use them for training until one (1) percent remains and then dispose of them at zero (0) percent life. Continued use at zero (0) percent could cause damage to the X26E. To replace the DPM:

- Ensure safety switch is down.
- Remove cartridge, if loaded.
- Depress DPM release.
- Remove and replace DPM.

9-53. When the DPM gets below zero (0) percent, the lithium energy cells are going dead. At this point, the power level will drop below the minimum level at which the microprocessor will run. This is called a brown-out. It is similar to unplugging your desk top computer from the wall without shutting it down properly.

9-54. When a DPM is replaced with a new DPM that contains a newer update of software than the current version in the X26E, an upgrade programming will occur.

9-55. A "P" will be displayed in the CID during the upgrade programming. This will take about 45 seconds. After upgrade programming has been completed, the unit will start the boot up sequence.

9-56. Removal of the DPM during the "P" state (programming) will corrupt the software. If this occurs, an "E" or "H" will be displayed. The X26E must then be returned to supply to be turned in for reprogramming.

PERFORM X26E FUNCTIONS CHECK

9-57. Do not point the X26E at others during the functions check. Keep the X26E pointed in a safe direction. Do not shine the X26E's laser at the eyes of others.

9-58. Perform X26E functions check prior to each mission to ensure that the device is fully operational. The functions check conditions components for regular use. Functions check (sparking) should not last longer than one (1) second.

> **WARNING:** The following exercises will be performed without a cartridge. The X26E will "FIRE" its electrical energy with or without a cartridge if the safety is off and the trigger is pulled. The X26E will cycle, spark, and expose you to unintended EMD and/or stun effect if you do not handle it properly. Keep hands, fingers, and any other parts of your body away from the contacts when the safety is off or when the X26E is discharging.

Functions Check Demonstration

9-59. Once the instructor completes the X26E TASER lesson, students observe a demonstration of PMCS and functions check procedures.

9-60. Perform X26E functions check using these procedures:

- Ensure the safety switch is in the down (SAFE) position.
- Remove the cartridge.
- Point the X26E in a safe direction and place the safety in the up (ARMED).
- Check the remaining battery life percentage on the CID. Replace the DPM if the percentage is less than 20 percent.
- Pull the trigger and perform a spark test for one (1) second.
- Look for visible spark between the electrodes and listen for a rapid spark rate. A full five (5)-second cycle is not required.
- Place the safety switch in the down (SAFE) position.
- Replace the expended cartridge, if present.
- If the functions check is not satisfactory, insert a new DPM and repeat the functions check. If the functions check is still not satisfactory, return the X26E to supply.

X26E FIRING PROCEDURES

9-61. Observe standard sidearm safety guidelines. Aim the X26E like a standard firearm at center of mass, ensuring that the weapon is held upright for a vertical target. The laser designates the point of probe impact within three inches at 13 feet. Be aware that the X26E fires its probes in line with an 8 degree spread. The spread between probes increases the further you get from your target with the probes separating one foot (.3m) for every 7 feet (2.1m) they travel. A minimum of 4 inches of probe spread is required for desired EMD effect.

9-62. The electric circuit is completed when both barbs hit the target. A 5-second cycle of electrical pulses begins and is sent through the wires into the target. The cycle can be stopped by moving the safety lever down or can be restarted by pulling the trigger again.

ENGAGE TARGETS WITH THE X26E

9-63. Aim at center of mass of the target. If the weapon is canted or held at an angle to the target area; the effectiveness of the cartridge to propel the probes properly will be decreased, which increases the probability of a miss.

9-64. Keep in mind that very close range results in high accuracy but less probe spread and a much greater potential for operator injury. Longer range provides increased operator safety and increased probe spread but increases the likelihood of a missed probe due to the subject torso size.

Chapter 9

9-65. An operator could misjudge maximum range and be slightly out of range. Moreover, the subject requires slack in the wires to fall down; therefore, if the subject is shot while running at maximum range, the operator must keep pace with the subject, as the running momentum of the subject will break the X26E wires. A range of seven (7)–15 feet provides the best overall compromise between accuracy and operator safety.

9-66. Longer distances (up to 25 and 35 feet) provide a wider probe spread (more effective) and a potential increase in operator safety but increase the risk of one or both probes missing the subject.

9-67. Probes are effective from point blank to 21 or 25 feet (depending on the type of cartridge used). Optimum range is seven (7) to 15 feet from target for probe spread, operator safety, and accuracy. The greater the probe spread, the greater the potential for increased effectiveness. If possible, a minimum of a four (4)-inch probe spread is desired.

9-68. The probes travel a path along a vertical plane aligned with the sights. One probe travels toward the point of aim; the other travels below the point of aim. Rotating the X26E off "plum" with the target can place the bottom probe off the target. Hold the X26E vertical in relation to the target.

9-69. If possible, fire the X26E at the subject's back for the following reasons:

- Normally, less interference from clothing, and clothing fits tighter.
- Surprise factor.
- Stronger muscles, increased effect.
- Reduces face, throat, or groin injuries.

Misfire Procedures

9-70. If the cartridge does not fire, it may fire after a delay. Even though the cartridge did not fire during the first pulse, there are 15–20 pulses-per-second, and any of these pulses may discharge the cartridge.

9-71. Make sure that the weapon is aimed at the intended target until the weapon is put in SAFE mode. If you aim the weapon off target while the X26E is still cycling, it could discharge the cartridge and hit an unintended target. If air cartridge is a "dud," (misfire) keep the X26E aimed on target and place it on SAFE. Discard the "dud" cartridge immediately. Reload the X26E with a new cartridge and re-engage target. Do not attempt to reuse a misfired cartridge. Immediately notify TASER International, provide the serial number, and return it.

EMD VERSUS STUN

9-72. The EMD weapons send out short-duration, high-voltage electrical waves that overpower the normal electrical signals within the nerve fibers, causing the body to "lock up." Stun weapons jam the central nervous system with electrical noise that affects the sensory nervous system, creating pain compliance. The X26E is both an EMD and stun weapon.

Drive Stun Application

9-73. Aggressively drive the X26E center mass of large muscle groups. Avoid the face, throat, and groin. Concentrate on the following target areas while pulling the trigger:

- Shoulder—Chest.
- Forearm.
- Thigh.
- Calf.

9-74. The X26E will always produce a charge when activated, even if an unfired cartridge is present. This allows the X26E to function in the stun mode immediately as a backup weapon without having to remove the expended cartridge. To use drive stun without firing probes, simply remove the live cartridge.

9-75. The drive stun mode only affects the sensory nervous system making it a pain compliance weapon that will not cause EMD. It is important to note that pain compliance does not equate to electro-muscular incapacitation. Some key points to remember when employing the X26E in drive stun mode:

- If the drive stun is not effective, evaluate location of drive stun and change target area.
- A drive stun with a live cartridge is possible.
- There is no over penetration from the cartridge being fired at close proximity to the subject's body.
- Cartridge might not deploy when in direct contact with subject but will still have drive stun effect.

PROBE REMOVAL

9-76. The preferred method to remove the probe from tissue is to stabilize the flesh with one hand and then to firmly and quickly pull the probe free. Make sure that the stabilizing hand is at least several inches away from the probe impact site. Some agencies have reported that operators have accidentally "raked" the barb across their own hand while removing a probe, breaking their own skin with a barb that is contaminated with the blood of the subject. It is critically important to stress the need to exercise extreme care with any object that has been exposed to body fluids. Once used, probes should be treated as bio-hazard.

9-77. Special medical procedures may be required for sensitive areas such as the eyes, groin, breast, etc.

9-78. When removing probes, check the probe body and ensure that the probe is intact and that the straightened barb is still attached to the probe body. There have been few reported cases in which the probe was removed from the subject but the straightened barb pulled free of the body of the probe and remained in the skin of the subject. Needle nose pliers or medical hemostats may be required to remove barbs from a subject. There have also been few reports of the barbed tip breaking off and remaining under the skin of the subject. In this instance, removal by medical staff is recommended.

DOWNLOAD USAGE DATA

9-79. The internal time clock of X26E may gain or lose several seconds or minutes a month. How much time is gained or lost is based in large part on the operating levels of several other components in the X26E. If the X26E goes several months or years without a download, the internal clock could be off by over an hour. This would result in inaccurate download information. By conducting a periodic download, the internal clock is reset to the download computer's time. Therefore, it is critical that the download computer is set to the correct time and time zone. The periodic download is a recommendation only. Refer to TB 9-1025-211-10.

9-80. The X26E Universal Serial Bus (USB) Data port: time, date, duration, temperature, battery status of each firing (over 1,500), connection protected inside DPM slot, encrypted data files, date range downloads, USB plug and play. The Commander must decide before training begins, who will download, when they will download, and where the records will be kept.

9-81. To download the X26E use the data port-download cable and follow the step by step instructions in the Data Port Operator Manual (V.15.6). Changes and updates to the manual can be viewed at www.TASER.com.

This page intentionally left blank.

Chapter 10

Employ Vehicle Barriers and Arresting Devices

OVERVIEW

10-1. Moving vehicles carrying large weights of explosives pose a tremendous threat. In volatile high-threat environments nonlethal vehicle barriers and vehicle arresting devices can stop threat vehicles before they can reach a checkpoint or access control facility.

10-2. This chapter provides general guidance for the application of nonlethal vehicle barriers and vehicle arresting devices that are currently being used by U.S. Army personnel world-wide. These nonlethal items include the Portable Vehicle Arresting Barrier (PVAB), the Vehicle Lightweight Arresting Device (VLAD), and other vehicle arresting devices such as Caltrops and Magnum Spike Systems. These and other useful nonlethal devices are all components of the Checkpoint Operations Mission Module. Other components of the Checkpoint Operations Mission Module are described in Appendix E.

PORTABLE VEHICLE ARRESTING BARRIER SPECIFICATIONS

10-3. The PVAB can be placed on any flat surface and used in any environment (see *TM 5-4240-342-12&P*). It is designed for vehicle denial and can successfully stop vehicles weighing up to 7,500 lbs. at speeds up to 45 mph within 200 feet. The standby mode of PVAB allows normal traffic to pass through regardless of volume without damaging the PVAB. The PVAB is used as a bi-directional system capable of one-lane (15 feet minimum) or two-lane (24 feet maximum) coverage and is remotely activated up to 300 ft. away. The PVAB NSN is 4240-01-469-6122.

10-4. The PVAB is most useful at access control points, deliberate checkpoints, and at the perimeter of controlled access areas. It can be anchored to alternate objects like guardrails or supports, bridge abutments, trees, or a parked vehicle such as a high mobility multipurpose wheeled vehicle. The PVAB can be placed in operation in less than one hour by two trained personnel. PVAB training is included in the Identify Arresting Vehicle Devices Lesson Plan available with the MMTSP.

10-5. The PVAB system consist of various components that include the following:

- Bump module assembly.
- Brake box.
- Anchor sets (A and B).
- Erector assembly.
- Control system.
- Net assembly.

PVAB SITE SELECTION

10-6. Ensure the area selected is large enough to accommodate an exclusion zone since captured vehicles may veer or components may fly off. In order to prevent possible inadvertent activation, do not place the control box where water may build up and flood the components.

10-7. Prior to installing, the roadway site should be free of potholes, large cracks and have dirt or gravel shoulders that are fairly level. The roadway should be fairly straight for 220 ft. in either direction of the PVAB system (or at least have no major obstructions for that distance) and 15-24 feet wide.

Chapter 10

10-8. The site selected must also be able to accommodate a hazard zone. The hazard zone should be 440 ft. by 84 ft. to prevent personnel injury when captured vehicles veer or if components of the system should break and fly. Figure 10-1 illustrates site requirements for proper installation and operation of the PVAB.

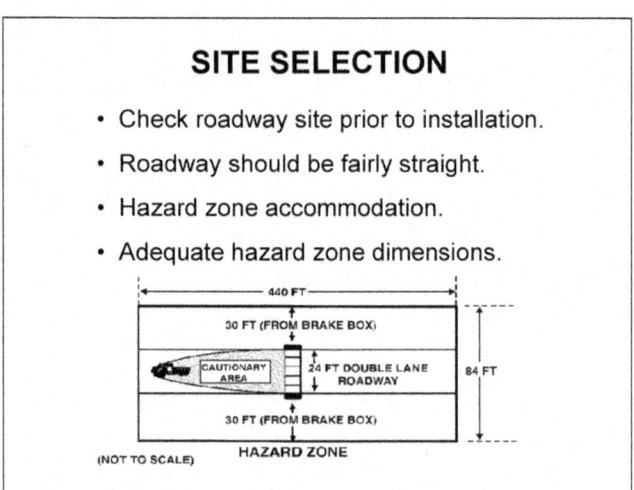

Figure 10-1. Site Requirements for PVAB

BUMP MODULE ASSEMBLY

10-9. Starting with the left bump end module (dog-bone end), place seven (7) bump modules and right bump end module across roadway, with top door hinges towards the expected primary direction of capture.

10-10. Face oncoming traffic. Starting with the left bump end module (dog-bone end) and working to the right, assemble modules by laying the slotted end over the dog-bone end of one bump module and firmly stepping on the two (2) modules to join them together. Ensure modules are flush when assembled. Continue this step to assemble all bump modules needed for road width, ending with the right bump end module (slotted end). Figure 10-2 illustrates this procedure.

Employ Vehicle Barriers and Arresting Devices

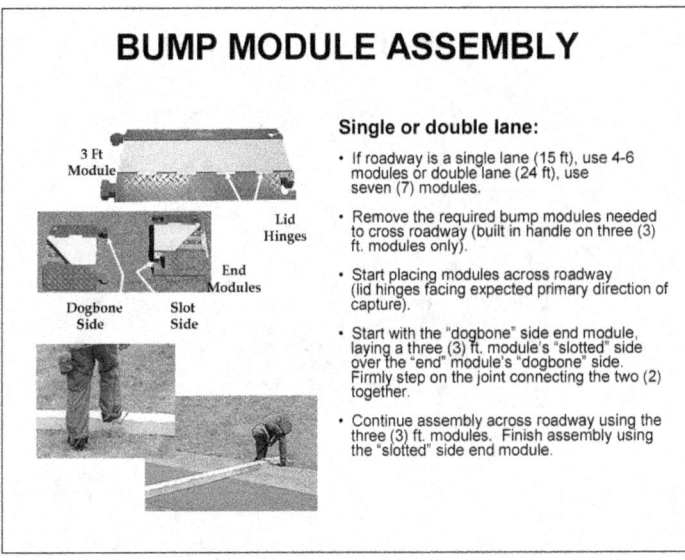

Figure 10-2. Bump Module Assembly

BRAKE BOX INSTALLATION

10-11. With the brake line slider positioned towards the road, place a brake box on each side of the road. The boxes should line up with each bump end module.

10-12. To connect a brake box to bump end module, lift upside of brake box next to bump-end module and ensure that holes on bottom of brake box align with alignment pins on stabilizer plate of bump-end module. Repeat for second brake box.

10-13. Ensure brake box is on each side of roadway. Position brake box so brake webbing (attached to "D" ring) faces end bump modules. Attach brake boxes to end modules by aligning holes under brake box onto the pins on stabilizer plate of end module. Figure 10-3 illustrates the brake box installation procedures.

Chapter 10

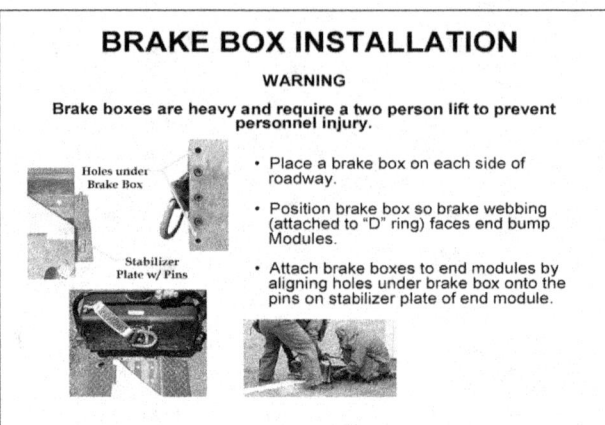

Figure 10-3 Brake Box Installation

ANCHOR INSTALLATION

10-14. Anchoring techniques vary according to soil type. The procedures described here are for normal soil. Use this method when the soil conditions allow the anchor plate to be buried 24 inches and soil can be easily re-compacted. Normal soil generally has enough moisture and clay content to allow the soil to be compacted or tamped. This soil would ordinarily be strong enough to drive a truck over (during dry conditions) without breaking through the surface and getting stuck. This soil is generally dark when moist. Refer to *TM 5-4240-342-12&P* for anchor instructions in other soil types. Installation under normal conditions is typically two hours using a two-person team.

10-15. Any anchoring method used may present a tripping hazard due to exposed wires and cables. Use extreme care to keep any exposed anchor cables covered with soil. Coil all wires and cables and keep them out of the foot traffic areas. If this is not possible, mark or flag the cables with a prominent material (police barrier plastic tape). If a long section of the 300 ft. pendant cable must be used, conceal the cable as much as possible. When digging the holes, avoid getting dirt on the brake boxes, since excessive amounts of dirt or other materials may affect the performance of the unit. The following equipment (not included in the NLCS may be needed:

- 3-lb. Sledge hammer (short handle) and 8 to 12 pound Sledge hammer (long handle). (NSN 5120-00-367-0250, 3 lb.) (NSN 5120-00-900-6096, 8 lb.).
- Arctic Tent Pins (for frozen soil conditions). (NSN 8340-00-823-7451).
- Steel Fence Post/Pickets for disturbed or extremely sandy desert. (NSN 5660-00-270-1587, 5 ft. long) (NSN 5660-00-270-1510, 6 ft. long) (NSN 5660-00-268-5411, 8 ft. long).
- Barrel, 55-gal plastic (preferred) or metal open end drums. (NSN 8110-01-343-1697, plastic) (NSN 8110-01-432-1207, steel).
- Picket pounders (locally fabricated).
- Additional items that you may need at initial set up are a cutting torch when using a metal 55-gallon drum, mattock, shovel/entrenching tool, and a Loc-tite.

10-16. For normal soil, the four 16-inch anchor-plate assemblies supplied with the M1 PVAB system should be installed following this procedure.

10-17. Attach loop of anchor cable to brake-box-eyebolt using provided screw pin shackle. Ensure screw pin is fully engaged but no more than one (1) finger tight. To determine the anchor location, grab anchor plate and stretch anchor cable away from brake box at a 30-degree angle from the roadway. Lay anchor plate on the ground. The location of the crimped ring on the anchor cable marks the center of the hole.3

10-18. Find the crimped ring two (2) ft. from the anchor plate. Mark this location and move anchor assembly out of the way in order to dig the hole for the anchor. When digging the holes, avoid getting dirt on the brake boxes, since excessive amounts of dirt or other materials may affect the performance of the unit.

10-19. Dig a hole at least 24 inches deep and approximately 24 inches in diameter. If the dirt is soft and easy to dig, the anchor should be buried a few inches deeper. Stack the dirt up fairly close to the hole, as it will be reused when burying the anchors.

10-20. The anchor cable has a crimped ring at 24 inches for proper depth. The hole is deep enough when the crimped ring of the anchor cable is even with or slightly below the surface level of the ground when the anchor cable is held vertically.

10-21. Loosen screw pin in screw pin shackle and temporarily remove anchor cable from brake-box eyebolt. Then finger-tighten the screw pin. Place anchor plate flat in bottom of the hole and tap down.

10-22. Cover the anchor plate with approximately 1/3 of the dirt previously removed. Tap the dirt down while holding anchoring cable vertically. Fill the hole 2/3 full and tap down. Reconnect anchor cable to brake-box eyebolt using provided screw pin shackle. If the cable is not reconnected now, there is a risk that the cable will not reach the brake box.

10-23. Fill in the rest of the hole with dirt and tap down. The cable should be partly covered with dirt. When water is available, wet down backfill to settle the dirt. If a vehicle is available, run vehicle over the hole to tap down further.

10-24. Repeat these steps for second anchor on the hinge side of the bump assembly and for the two (2) anchors on the other side of the bump assembly. Figure 10-4 illustrates the anchor installation procedures.

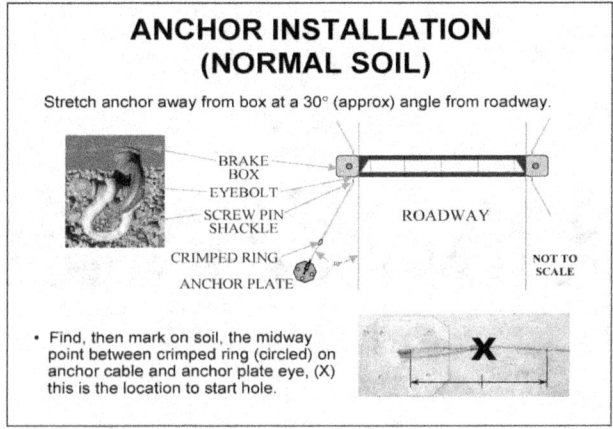

Figure 10-4. Anchor Installation

ERECTOR ASSEMBLY

10-25. Ensure air hose fitting on erector assembly does not get caught in brake-box mounting cup. Insert base of erector assembly into mounting cup on top of brake box, while aligning air-hose fitting into slot. Push down until firmly seated. Repeat for second erector assembly. Open all bump-module lids. If a bump-module lid falls off, align lid hinge over hinge slots in bump assembly and stamp down on hinge to connect the two pieces. Connector end must be on the outside of the bump-end module. Make sure excess length of 30-foot interconnect control cable is on the same side as control box.

10-26. Determine where control box will be located. Starting at the side opposite the control box, push connector end of 30-foot interconnect control cable down through top hole and out through side hole in the bump end module. Align the end of the 30-foot interconnect control cable with the erector assembly connection. Push both together to connect. Lay the 30-foot inter-connect control cable into channel one (1).

10-27. Push the other connector end of the 30-foot inter-connect control cable through two (2) holes in the other bump end module and over the top of the anchor cable. Connect one (1) end of the 6-foot inter-connect control cable to the erector assembly connection that is on the same side as the control box. Use steady and constant strokes at a moderate pace when using the air pump or it may burn up.

10-28. Unscrew the cap from the fill valve on the air-bottle assembly. Set the cap aside. Screw the air pump valve onto the fill value. Use air pump to pressurize each air bottle so that the valve assembly gauge reads a minimum of 120 psi, but not more than 130 psi. When checking air pressure, always use the gauge on the valve assembly.

10-29. Unscrew the air-pump valve from the valve assembly and remove the air pump. Place the screw cap back on and fill valve. If pressure in the air pump needs releasing, press the valve stem pin. Pump the air pump a few times to cool it.

INSTALL CONTROL SYSTEM

10-30. Excess cable length will allow the control box to be positioned a safe distance (approximately 6 ft.) away from the brake box. Align free end of the 30-foot and six-foot inter-connect cables with a 3-pin connector on the control box. Push down to connect. Both of the 3-pin connectors on the control box are the same.

10-31. Align one end of the 300-foot pendant control cable with connector on the end of the pendant. Push together to connect. Align free end of 300-foot pendant control cable with two-pin connector on control box and push down to connect it to the two-pin connector.

10-32. If the 300-foot control cable has been unrolled from the cable reel to allow the pendant to be at a distance from the control box, bury any excess length to prevent a tripping hazard. If this is not possible, mark or flag the cable with a prominent material (police barrier plastic tape/engineer tape).

> **Warning:** Ensure all personnel are cleared from the area over the bump modules and erector assemblies before activating the system.

10-33. Flip up the red switch cover on the pendant. Press and release FIRE switch. The erectors should be fully extended in approximately two (2) seconds. Close red switch cover on pendant. If erector assemblies do not function, go to the troubleshooting procedures in *TM 5-4240-342-12&P*.

NET ASSEMBLY

10-34. Two (2) people are required to install the net assembly. Place plastic bag containing the net in the middle of the roadway, on the side of bump assembly and away from oncoming traffic. The net is pre-packed in a plastic bag with two (1) net ends nearest the bag opening. Each person grabs one (1) of the net ends and while walking away from each other pulls the net out of the bag, and stretches the net across the road. This will avoid tangling the webbing.

Employ Vehicle Barriers and Arresting Devices

10-35. If net is tangled find the top of the horizontal-webbing where the center vertical-webbing joins the horizontal and diagonal pieces. Follow the horizontal webbing out to the ends, twisting it as needed to lie flat from end-to-end. Stretch out the net assembly on the roadway. The net should be laid out as follows:

- Place the (black) rear-capture line closet to the bump (the black rear-capture line is not attached to any webbing and has black tape on ends).
- The (white) bottom-lateral line is placed next (the white bottom-lateral line is stitched to the three-inch webbing).
- The (pink) front-capture line is placed farthest away from the bump (the webbing at the top of the net is looped around the front capture line).

10-36. Figure 10-5 illustrates the net assembly.

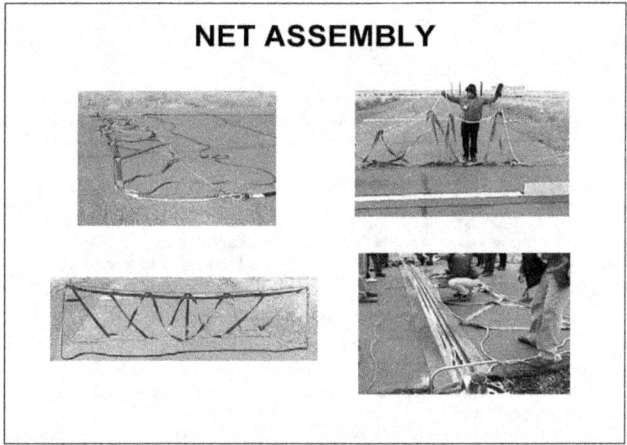

Figure 10-5. Net Assembly

INSTALL LIFT STRINGS

10-37. This procedure requires the erectors to be extended. Without lift strings attached, the erector will not raise the net upon activation.

10-38. Remove one lift string from plastic bag. Feed one looped end of lift string between the upper-retention band and erector assembly on the same side as the bump modules on road. Take the other end of the lift string and place it through the loop. Pull tight to secure lift to upper-retention band on erector assembly.

10-39. Feed the free end of the lift string through the quick link at top of the erector. Attach the loop of free end of lift string to the large quick link at the top of the net by loosening the quick link nut all the way and inserting the loop of lift string. Tighten the quick-link-nut finger tight. Repeat steps to install a lift string on the other end of the net.

10-40. To attach the net assembly to brake-line webbing ensure that five (5) feet of black rear capture line (approximately double arm span) extends beyond the ratchet block. If it is difficult to use your fingers to open the ratchet cam, use an available tool, stick or tent pin.

10-41. If there is not five (5) feet, push in on the ratchet cam, to release the tension while pulling out on the (black) rear-capture line. Release the cam lock back into place. The Rear-capture line should not be able to be pulled in or out, if cam is locked.

5 November 2009　　　　　　　　　　　　TC 3-19.5　　　　　　　　　　　　10-7

10-42. Grab the center of (black) the rear-capture line and place it in the center of channel two (2) (middle channel) of bump assembly. Working from the center outward, continue to place the rest of (black) the rear-capture line in channel two (2) of the bump assembly. Ensure the center of the rear-capture line remains in center of the bump assembly. Notice that all three (3) connection points (disposable tension link (DTL), rear-capture line and brake-line slider) have black markings.

10-43. Remove a DTL (approximately 10-inch long black and orange rope with black marking) from plastic bag. Remove screw-pin from screw pin shackle on brake line slider. Set screw pin aside. Place the end loop of (black) the rear-capture line and one end of the DTL onto the screw-pin shackle on the brake-line slider. Place screw pin back in screw pin shackle and finger-tighten.

10-44. Loosen the screw pin in the screw pin shackle connected to the brake box eyebolt on the hinge side of the bump and attach the free end of the DTL. The Screw pin-shackle connected to the brake-box eyebolt should now hold one (1) end of the DTL and the anchor cable. Repeat brake box steps for the opposite brake box.

10-45. Run the looped end of (white) bottom lateral line under (black) the rear-capture line, then through the D-ring on the brake webbing. (It does not matter which direction the lopped end is placed through D-rings). Ensure lines are not tangled.

10-46. Loosen the quick-link nut at the top of the net and attached looped end of (white) the bottom lateral line. Finger-tighten quick-link nut. Ensure that the (pink) front-capture line, lift string, and net do not fall out of quick link. It should not hold the (pink) front-capture line, lift string, net, and (white) bottom-lateral line.

PACK NET AND FRONT-CAPTURE LINE

10-47. Collapse erectors and verify FIRE switch is off and red-switch cover on pendant is closed. For one-person operation follow these procedures; Push in vent-valve button to release some pressure. Release button and then push down on top of erector. Repeat procedure until erector is fully collapsed. Repeat this step for second erector.

10-48. For two-person operation; Operator #1 pushes in the vent-valve to release pressure while Operator #2 pushes down on top of erector until it is fully collapsed. Repeat this step for the second erector.

10-49. Ensure the (black) rear-capture line is in channel 2. Place center of (white) bottom-lateral line (with wet-webbing) into the center of channel 3 of the bump assembly. Continue to place the rest of the bottom-lateral line (with net webbing) into channel 3. Ensure center of bottom lateral line remains in the center of the bump assembly. Pull ratchet block and large quick link over to each bump-end module. Do not tangle excess line. The (pink) front-capture line must sit on top of everything in channel 3 to assure proper deployment. Ensure the center of the webbing always remains in the center of the bump assembly while packing the net.

10-50. Starting at the center of the bump assembly and working outward, start packing net webbing into channel 3. When all of the webbing is in channel 3, place the center vertical part of webbing with (pink) front-capture line on top of the webbing already packed in channel 3. Ensure the (pink) front-capture line sits on top of everything in channel 3 and lines are not tangled. No webbing or lines should be between the lid and channel wall and sticking out of the lid.

10-51. To prevent webbing and lines from slipping out of channel 3, starting at center of bump assembly and working outward, close center bump module lid and the two (2) module lids on each side of the center module as packing of webbing is completed. Before closing the last full-length bump module lid on each end of the bump assembly, fold over excess capture line and place excess line into channel 2.

10-52. Ensure that the ratchet blocks and the quick link are placed flat in both bump-end and modules so those lids will close. Lift string must lie on top of all components in bump-end module to ensure proper net deployment.

10-53. Place ratchet block and quick link on top of excess capture line so that they lay flat in both bump-end modules. Place lift string on top of all components in both bump-end modules. Ensure DTL's and lift string exit through diagonal cutout on the lid of both bump-end modules. Close both bump-end module lids. Pull slack out of lift strings, tucking excess under lids.

Employ Vehicle Barriers and Arresting Devices

10-54. Figures 10-6 and 10-7 illustrate the pack net procedures.

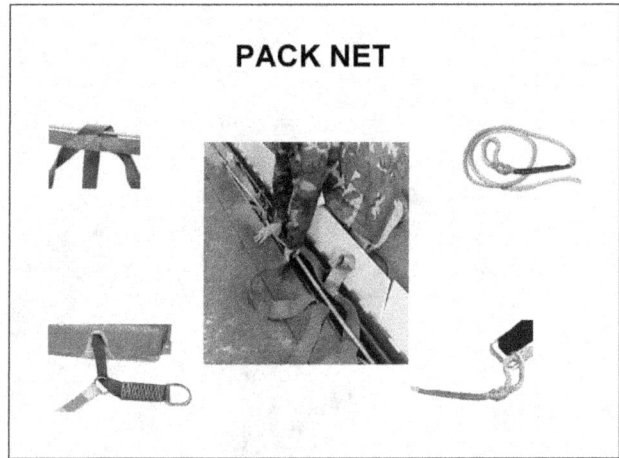

Figure 10-6. Pack Net

Figure 10-7. Pack Net Continued.

10-55. Users of the PVAB system should now conduct a systems test using the detailed procedures described in the PVAB technical manual. Operators must follow the activation, disassembly, and safety requirements prescribed in the technical manual. See *TM 5-4240-342-12&P*.

Chapter 10

VEHICLE LIGHT ARRESTING DEVICE SPECIFICATIONS

10-56. The Vehicle Lightweight Arresting Device (VLAD) is a counter-materiel, non-lethal vehicle stopping device. The VLAD provides a man-transportable and quickly pre-emplaced means of stopping several different classes of wheeled vehicles. The device is equipped with a unique barbed spike system that can be used in various scenarios such as at checkpoints, roadblocks, and dismount points to stop and ensnare cars and trucks. User of the VLAD system should refer to TM 5-4240-536-10 for technical guidance on installation, safety, and operation. The VLAD NSN is 4240-01-518-4626. Figure 10-8 shows the VLAD system packed ready for transport.

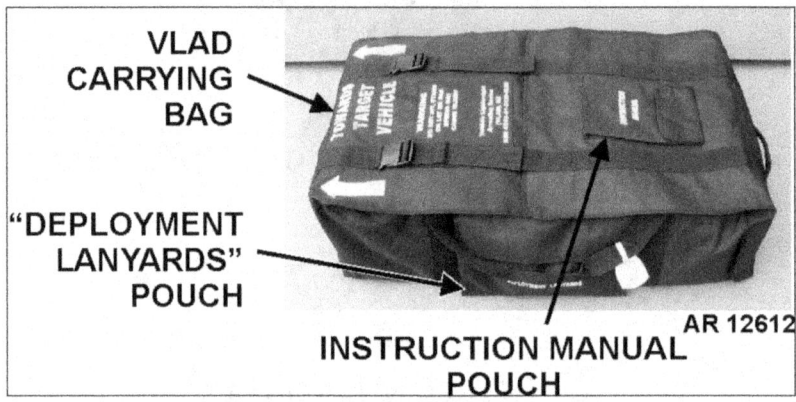

Figure 10-8. VLAD System Packed for Transport

10-57. The VLAD can be rapidly deployed manually, either by laying it out on the road as shown in Figure 10-9, or by placing it on the side of the road and simply pulling the lanyards to draw it across when needed. Once the oncoming vehicle's tires pass over the spikes, the entire system wraps around one or both front tires and axle, forcing the vehicle to slow down and come to a complete stop.

10-58. The net is manufactured from high modulus polyethylene fiber called Dyneema. This material is extremely lightweight and strong and is capable of binding to the axle without slipping or tearing. The netting configuration allows for kinetic energy absorption, and the spikes are machined from high carbon steel to allow them to pierce most tires.

10-59. There is no overt concern about significant injury or lethality potential with proper employment of the VLAD. This device is capable of stopping vehicles with minimal damage and minimal harm to the vehicle occupants and surrounding personnel. Vehicle engagements with the VLAD may result in minor injuries to vehicle occupants not wearing a seatbelt. The evaluation was conducted using similar modeling and simulation and an injury assessment report. Figure 10-9 illustrates the VLAD system deployed across a road.

Employ Vehicle Barriers and Arresting Devices

Figure 10-9. VLAD Positioned Across Roadway

VLAD SITE SELECTION

10-60. Since a fast moving vehicle will have a much higher chance of evading the net device, the site selected should be just after a natural slowing feature to maximize the chance of a successful mission. Typical stopping distances for medium passenger vehicles and light trucks (less than 2 tons) are shown in Table 10-1.

Table 10-1. Vehicle Stopping Distances

VEHICLE SPEED	TYPICAL STOPPING DISTANCE
30 MPH (48 km/h)	82 ft. (25m)
40 MPH (64 km/h)	118 ft. (36m)
50 MPH (80 km/h)	180 ft. (55m)

10-61. In dry conditions there should be a minimum of 330 ft. (approximately 100m). In wet conditions the distance should be doubled.

VLAD RAPID DEPLOYMENT

10-62. Once a deployment site has been selected, determine the most likely direction of the threat vehicle. Using two (2) people, each person grabs a carrying strap or each person grabs a handle on the end of the VLAD carrying bag. Place carrying bag just off the roadway with arrows marked "Towards Target Vehicle" pointing towards threat vehicle approach and the pouch marked "Deployment Lanyards" at curb/roadside.

10-63. Using thumb and index finger on each side of the buckle, squeeze both buckle releases together to unclip outside strap on carrying bag. Repeat for second outside strap. Pull open outside flaps #1 and #2. Pull open inside flaps #1 and #2. One (1) end of carrying bag is closed with hook and pile closures. To open that end of the carrying bag, pull the hook and pile closures apart and lay it flat.

10-64. Using two (2) people, one (1) person carefully places both hands under the entire spiked section of the folded net and lifts it up while the other person pulls the carrying bag out of the way. Set the carrying bag aside for possible repacking.

10-65. One (1) person lifts the net device while the other person pulls the carrying bag out of the way. Ensure you save the carrying bag for repacking should the VLAD not be deployed. One (1) person carefully holds the spiked net in place as the other person grabs the two (2) yellow tabs on the other end of the net and walks away pulling the net until it is unfolded. Net should be parallel with the roadside.

10-66. Using two (2) people, each person grabs a red deployment handle and pulls the net device across the road approximately at a 45 degree angle (See illustration below). The barbed spikes on the leading edge of the net should be upright and net should be pulled tight. Do not remove plastic covers over barbed spikes. They are designed to crush down and expose the spikes when a tire runs over them.

10-67. If necessary and time allows, reposition the net and make sure the net is as flat as possible to avoid being detected by threat vehicle. Do not step on barbed spikes. Stand clear of the area.

10-68. If the VLAD was deployed but not used to stop a vehicle, repack the net device. Refer to the technical manual for repacking instructions.

VLAD LANYARD DEPLOYMENT

10-69. Lanyard deployment procedures are the same as rapid deployment with the additional steps of removing two (2) plastic bags from the pouch on the outside of the carrying bag marked "Deployment Lanyards". Remove two (2) plastic anchor pegs, two (2) anchor lanyards, and two (2) deployment lanyards from plastic bags. Place empty plastic bags back in pouch for possible re-use. Set carrying bag aside for possible repacking.

10-70. To anchor the net device, determine if there are soft edges at the side of the road. If there are use the two plastic anchor pegs to secure the net device. Insert a plastic anchor peg through the grommet in each of the two (2) white straps on the end of the net. Push plastic anchor pegs into the ground. Do not hammer in or force in too deeply.

10-71. If there are no soft edges at the side of the road, Clip an anchor lanyard/strap to the buckle at the end of the white strap at each side of the net. Run one (1) of the anchor lanyards out to a softer area at a 45 degree angle (approx.). Insert an anchor peg into one (1) of the three (3) pre-sewn anchor peg loops on the anchor lanyard. Using hands, push anchor peg into ground ensuring loop is securely under the anchor peg hooks on the top of the anchor peg. Do not hammer in or force in too deeply. Repeat at other side of net.

10-72. Clip a deployment lanyard to the buckle at the end of the two red deployment handles on the net device. Using two (2) people, each person grabs an end of a deployment lanyard and walks it to the opposite side of the road at a 45 degree angle (approx.). Do not unfold the net at this time.

10-73. From a hidden vantage point, each person loosely holds the end of a deployment lanyard and awaits a target vehicle. DO NOT wrap end of deployment lanyard around hands or fingers.

10-74. Just before target vehicle approaches, both people quickly pull deployment lanyards across the road and immediately release the lanyards. Stand clear of area in case target vehicle makes evasive maneuvers or tries to avoid the VLAD.

10-75. Once the vehicle has stopped, quickly gain control of the occupants. Use caution when approaching the vehicle.

REMOVE NET FROM VEHICLE

10-76. Avoid unnecessary handling of the net device to prevent hand injury from barbed spikes. Wear leather gloves and do not step on barbed spikes on the leading edge of the net. Use a sharp knife to remove the bulk of the net from vehicle tires.

10-77. Dispose of any remaining netting to prevent unauthorized re-use. The vehicle will be incapacitated and must be recovered in accordance with local vehicle recovery procedures.

10-78. The VLAD is a one time use item. However, if the system was deployed but not used to stop a vehicle, perform the repacking instruction described in TM 5-4240-536-10.

RAPID DEPLOY VEHICLE DENIAL SYSTEMS

10-79. Quick deployment vehicle denial systems can be useful supplements for permanent or portable barriers or other vehicle arresting devices such as PVAB and VLAD. Devices such as the magnum spike system and caltrops are easily deployed and recovered and very effective against most threat vehicles.

MAGNUM SPIKE SYSTEM

10-80. The Magnum Spike System comes in a carrying case that deploys 160 uniquely designed spikes configured in rows along a 16 foot mat. The system comes with 45 feet of deployment cord on a reel with press button release and quick rewind. Spikes are designed for large air escape and engineered to stop and cut up all types and sizes of tires including run-flats. Clog-proof spikes will withstand multiple high speed impacts. Included are 50 replacement spikes with holder retainers that can be changed in the field with no tools required.

10-81. The system is simply unrolled from a safe distance across a designated area in front of a threat vehicle. Figure 10-10 illustrates the Magnum Spike System.

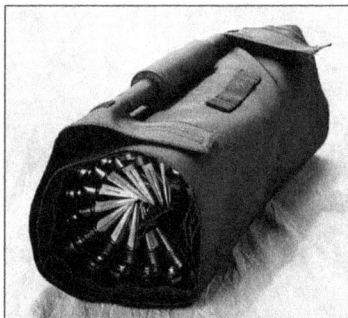

Figure 10-10. Magnum Spike System

CALTROPS

10-82. Caltrops can be hand-scattered and when employed, always land with the point up. They are "V" shape by design to ensure rapid tire deflation, incapacitating the target vehicle. Caltrops are used for vehicle denial, and they supplement permanent or portable barriers or other arresting devices like the PVAB or VLAD. Several Caltrops strung together can be used alone at a hasty checkpoint. Figure 10-11 illustrates typical caltrops.

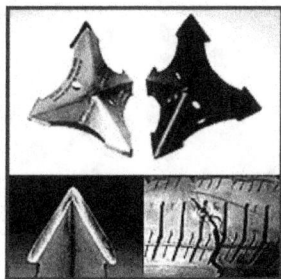

Figure 10-11. Caltrops

This page intentionally left blank.

Chapter 11

Employ Communication Devices

OVERVIEW

11-1. The techniques described in this chapter are used to employ the Phraselator, Voice Response Translator (VRT), and Magnetic Audio Device (MAD). The Phraselator and VRT translate English to another language, providing a means of communications between people who lack a common language. The MAD amplifies sound, which allows the user to talk to a multitude of people at safe distances.

PHRASELATOR SPECIFICATIONS

11-2. The Phraselator handheld translation device is a mobile, voice, and touch activated device that translates English phrases into other languages. When you speak a phrase, the device speaks the equivalent phrase in the language you choose. Phrases are predetermined by industry experts and recorded by native linguists to ensure accurate and successful communication. The Phraselator solves the problem of how to communicate important information in the absence of a human translator through a series of industry and culturally appropriate phrases that are preloaded into the device.

11-3. While the Phraselator is a one-way translation device (English to other languages), phrases are designed to prompt responses that can be conveyed using gestures or visual aids, for example: "Show me on the map" and "Raise your hand if you understand."

FEATURES AND CONTROLS

11-4. The diversity of phrases and languages available for the Phraselator makes it a powerful tool for anyone who needs to communicate with non-English speakers. The device is not intended to replace a human translator, but can facilitate immediate and necessary communication when a human translator is not available. Figure 11-1 illustrates the features and controls of the Phraselator.

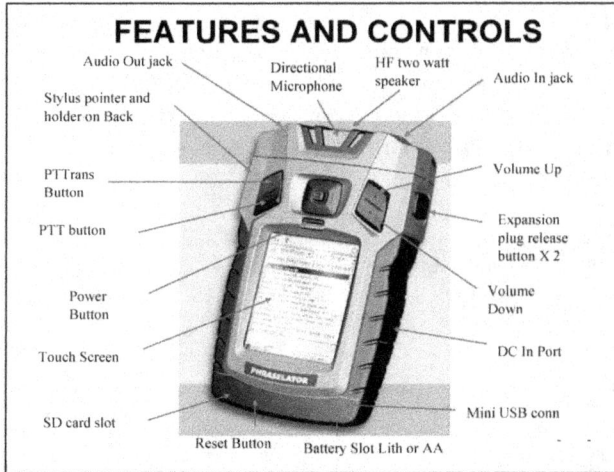

Figure 11-1. Phraselator Features and Controls

PHRASELATOR EMPLOYMENT CONSIDERATIONS

11-5. The Phraselator is an ideal communications tool for Soldiers who operate in remote locations and do not have access to a translator. Use the Phraselator for cordon and search, checkpoint/roadblock operations, information collection, or information dissemination.

11-6. The Phraselator is used to:

- Ask yes or no questions where the response can be given with gestures. For example, "Do you feel sick?" "Move your head like this for yes." "Move your head like this for no."
- Give orders or instructions. For example, "Do not move."
- Ask questions when the response can be to point to something. For example, "Show me where it hurts most."
- Convey information. For example, "We must evacuate the area."

OPERATE PHRASELATOR

11-7. Your Phraselator comes with an SD card, which contains version (v) 3.0 application software, modules, and depending on the SD card's size, extra room for new phrases and recordings. To install or change the SD card, follow these procedures:

- Be sure that you have exited (v) 3.0 application software. Do this by tapping the EXIT PHRASELATOR button under the OPTIONS tab.
- Turn the metal D-rings at the bottom of the Phraselator ¼ turn counterclockwise and then pull them out slightly to loosen the rubber cover. Remove the cover by peeling it away from the base.
- Inside is a slot labeled "SD Card." If an SD card is installed, press the card in slightly to unlock it.
- Firmly pull the card out from its slot. If the card doesn't easily release, don't force it. Try repeating pressing the card to unlock it.
- With the gold pins facing up and toward the Phraselator, place the SD card gently into the slot and push it into place. You should hear a soft click as it locks in.
- Replace the rubber cover.
- Push in and turn the metal D-rings ¼ turn clockwise to lock them in place.

TURNING ON THE PHRASELATOR

11-8. The POWER button is centered immediately above the screen. Press the POWER button to turn on the Phraselator.

- From the off power mode, you will automatically be taken to the screen from which you last exited the v3.0 application.
- If the Phraselator was in deep sleep mode, you will need to recalibrate your screen and (possibly) confirm the system date and time before the application will open.

SELECTING A MODULE

11-9. The Phraselator comes with Modules, or sets of phrases translated into various languages, that are customized to specific communication needs. Modules are loaded onto an SD card. If there is more than one Module on an SD card, you will need to load one when the v3.0 application starts. You can load additional Modules by following these procedures:

- Select the OPTIONS tab.
- If needed, select the MODULES sub-tab. The list of available Modules is displayed. If you have recently changed SD cards, tap on the REFRESH LIST button.

- Tap on the desired Module. If you are switching from one Module to another, a window asking you to verify your choice is displayed. Tap the Yes button.
- The new Module is loaded and you are taken to the PHRASES tab.

Note: If only one Module is on the installed SD card, that module will load by default when the device is turned on and you will be taken directly to the PHRASES tab.

SELECTING A LANGUAGE

11-10. You can change the output language of the Phraselator using the pull-down menu at the top of the screen or by using verbal commands to tell the device to switch languages. Only languages in the loaded Module are displayed in the menu. The default language selected is the first language listed under English.

Using the Stylus

11-11. Follow these procedures to use the Stylus:

- Tap the language pull-down menu at the top of the screen. The list of available languages is displayed.
- Tap or scroll down to the desired language.

Using Verbal Commands

11-12. Follow these steps when using verbal commands:

- Press and hold the PUSH TO TALK™ button.
- Speak the following command: "Phraselator change to {state the language you wish to choose}."
- When finished speaking the verbal command, release the PUSH TO TALK™ button and the language selection will change as requested.

SELECTING A CATEGORY

11-13. Categories within a Module are presented in a pull-down menu under the PHRASES tab. Categories appear with folder icons next to them and may contain subcategories. The ALL PHRASES category contains a listing of every phrase available in the particular Module being used.

11-14. If possible, select a category other than ALL PHRASES to improve accuracy and speed. The fewer phrases the Phraselator must search through to find the desired phrase, the better. The on-screen phrase list under the PHRASES tab shows all phrases available for the selected category.

Note: The P2 only recognizes phrases in the selected category. If you select a CATEGORY that does not contain the phrase you want to speak, the Phraselator will not be able to recognize that phrase.

Using the Stylus

11-15. Follow these procedures when using the stylus:

- Tap the down arrow of the CATEGORY pull-down menu.
- Tap the desired category. All phrases in the selected category are displayed in the phrase list under the PHRASES tab.

Note: You must go back to the top level before moving from one category to another.

Chapter 11

Using Verbal Commands

11-16. Use the following procedures when using verbal commands:

- Press and hold the PUSH TO TALK™ button.
- Speak the following command: "Phraselator change to {state the name of the category you wish to choose}."
- When finished speaking the verbal command, release the PUSH TO TALK™ button and the category selection will change as requested.

Note: You must go back to the top level before moving from one category to another.

TRANSLATING PHRASES

11-17. There are several ways to translate a phrase: Stylus and touch screen, voice recognition, or PUSH TO TRANSLATE button.

Note: There are a few special features associated with phrases in modules designed for the device's software application. For instance, a phrase can have an associated "Short Phrase" making the recognition easier and faster (for example, the short phrase "Read Miranda Rights" prompts the device to say the complete Miranda Rights), or prompt an Auto-Record feature. Short phrases and Auto-Record Phrases are indicated by unique phrase icons and are discussed in more detail later in this session.

Using the Stylus

11-18. Verify that the Phrases tab is active by tapping the PHRASE tab. Drag the stylus down the scroll bar or phrase list to find the desired phrase, or use the TOGGLE button to browse to the desired phrase. Tap the desired phrase. Verify that the device speaks the selected phrase in the selected language.

11-19. Tap the desired phrase. You may need to scroll through the phrase list to find the desired phrase. Do this by dragging the stylus down the scroll bar or phrase list, or move the TOGGLE button until you find the desired phrase. You can refer to the Module Reference Card™ to find the category for the desired phrase. The Phraselator speaks the phrase in the selected language.

Using Voice Recognition

11-20. Be sure you are on the PHRASES tab. If not, do one of the following:

- Tap the PHRASES tab.
- While pressing the PUSH TO TALK™ button, say, "Phraselator change to PHRASES tab."
- Use the TOGGLE button to navigate to the PHRASES tab.

11-21. Hold the Phraselator four to six inches from your mouth. Press down and hold the PUSH TO TALK™ button. Speak the desired phrase exactly as it is written on the phrase list or Module Reference Card™. Do not omit or change words.

11-22. When you finish speaking the phrase, release the PUSH TO TALK™ button. The Phraselator speaks the phrase in the selected language. If the Phraselator does not recognize a spoken phrase, you'll hear a short beep and see a "Phrase match not found" message on the screen. In this case, speak the phrase again.

Note: Only phrases shown in the phrase list under each category are included in your Module.

USING THE PUSH TO TRANSLATE™ BUTTON

11-23. The PUSH TO TRANSLATE™ button is used to repeat, translate, or verify a selected phrase. Text and audio verification modes are used with the PUSH TO TRANSLATE™ button to verify that the phrase you speak into the Phraselator is correctly recognized. There are two different modes you may use when using the PUSH TO TRANSLATE™ button: AUDIO VERIFICATION mode or TEXT VERIFICATION mode.

11-24. When the AUDIO VERIFICATION mode is enabled, if you use the PUSH TO TRANSLATE™ button to speak a phrase, the Phraselator will repeat the phrase you spoke in English before speaking the translation to ensure the correct phrase was recognized. You can then press the PUSH TO TRANSLATE™ button to translate the phrase into the selected language.

11-25. In the TEXT VERIFICATION mode, if you use the PUSH TO TALK™ button to speak a phrase, the Phraselator will display the phrase you spoke to ensure that the correct phrase was recognized. You can then press the PUSH TO TRANSLATE™ button to translate the phrase into the selected language.

11-26. Be sure you are on the PHRASES tab. If not, do one of the following:

- Tap the PHRASES tab.
- While pressing the PUSH TO TALK™ button, say: "Phraselator change to PHRASES."
- Use the TOGGLE button to navigate to the PHRASES tab.

11-27. Move the TOGGLE button up or down to find the desired phrase. When the phrase you want is highlighted, press the PUSH TO TRANSLATE™ button or press the center of the TOGGLE button. The Phraselator speaks the phrase in the selected language.

Using Short Phrases

11-28. Short phrases are essentially abbreviations of longer phrases. When you speak a short phrase, the Phraselator speaks the associated longer phrase in the selected language and shows English text for the complete phrase on the display, so you know what was translated. For example, the series of phrases used to describe the Phraselator could be triggered by a short phrase. Instead of saying, "This is a computer translator," then "I will use this machine to talk with you," and finally "The machine only works from my language to yours," you could say the short phrase, "Explain the Phraselator," and the Phraselator would say the complete series of phrases.

Using Auto-Record Phrases

11-29. The Auto-Record feature tells the Phraselator to automatically start recording after speaking a translated phrase. This is useful for phrases that require a response, as the response is recorded for later translation.

11-30. Tell the Phraselator to translate using any of the methods described. The Phraselator will speak the phrase then immediately start recording a response. When the person responding is done speaking, tap the STOP button.

Note: After stopping a recording, you can select PLAY to immediately play back the recording, or go to the RECORD tab to play, rename, delete, or move the recording.

11-31. When finished, select the X in the upper right corner to close the dialog. Once the dialog is closed you can find the recording under the RECORD tab. It will be named with the name of the phrase that prompted auto-record, and have the day's date assigned to it. If you have more than one recording for the same phrase on the same day, the recordings will be numbered Response 1, Response 2, and so on.

RECORDING

11-32. The Phraselator can create recordings for playback later. Use this capability to record yourself or responses from people you are communicating with for later translation.

11-33. Every 45 seconds of recording requires approximately one megabyte (MB) of storage. Depending on the size of your SD card, storage can be used up very quickly. Be mindful of this when using the record function and download recordings from the SD card often.

11-34. The Phraselator indicates at the bottom of the screen how much free space is available for recording on the SD card. When the maximum available storage is reached, the Phraselator will no longer record until more space is freed for additional recordings. You will be alerted with the following message when there is no more space for recordings on the SD card: "Recording has been stopped. The storage card may be full. You may want to delete some recordings and try again." Table 1-1 is provided to give approximate ratios of recording time to storage used.

Table 11-1. Approximate Ratios of Recording Time to Storage Used

RECORDING TIME	STORAGE USED
45 seconds	1 MB
2 minutes	2.7 MB
5 minutes	6.7 MB
8 minutes	10.7 MB
10 minutes	13.3 MB

Making a Recording

11-35. Be sure you are on the RECORD tab. If not, do one of the following:

- Tap the RECORD tab.
- While pressing the PUSH TO TALK™ button, say, "Phraselator change to RECORD tab."
- Use the TOGGLE button to navigate to the RECORD tab.

11-36. You may begin the recording two ways:

- Tap the RECORD button on the screen.
- Hold down the PUSH TO TALK™ button.

11-37. Begin speaking or have the person being recorded speak. Recording begins immediately. Depending on how you started the recording, do one of the following to stop the recording:

- Tap the STOP button.
- Release the PUSH TO TALK™ button.

11-38. An audio file is immediately displayed under the RECORD tab.

Playing and Deleting Recordings

11-39. Be sure you are on the RECORD tab. If not, do one of the following:

- Tap the RECORD tab.
- While pressing the PUSH TO TALK™ button, say, "Phraselator change to RECORD tab."
- Use the TOGGLE button to navigate to the RECORD tab.

11-40. Tap the desired audio file under the RECORD tab, or use the TOGGLE button to highlight it. Tap the PLAY button or press the TOGGLE button at the center to listen to the recording. Tap the STOP button to halt the playback.

11-41. Tap the DELETE button to delete the recording, or tap and hold your stylus to the touch screen and a menu will appear. Select DELETE RECORDING and then tap the YES button.

Renaming Recordings

11-42. Recordings you create from the RECORD tab are given a standard name including the time and date of the recording. Recordings made using the Auto-Record feature are named by the phrase that prompted the recording, followed with the response number if multiple responses are made in one day. While the recordings will be created with one of the above-named standards, the user may want to rename the recording to something more meaningful. The following directions describe how to rename a recording.

11-43. Be sure the device is on the RECORD tab. If not, do one of the following:

- Tap the RECORD tab.
- While pressing the PUSH TO TALK™ button, say, "Phraselator change to RECORD tab."
- Use the TOGGLE button to navigate to the RECORD tab.

11-44. Tap and hold the stylus on the desired audio file under the RECORD tab. Select RENAME RECORDING from the menu. The INPUT PANEL will appear.

11-45. Use the stylus to type the desired name for the audio file on the INPUT PANEL, and then tap the OK button at the top. If you do not wish to save the changes, tap the X instead.

Accessing Recordings (Audio Files)

11-46. Recordings are stored on the SD card with the Module(s). By default, recordings are stored as .wav files. They can be downloaded directly from the Phraselator to a personal computer (PC) with Microsoft ActiveSync and the supplied mini USB cable. To access recordings on your PC:

- Open WINDOWS EXPLORER on the PC (or EXPLORE when using ActiveSync).
- Double click the STORAGE CARD folder.
- Double click the PHRASELATOR folder.
- Double click the Module folder that was selected when the recording was created.
- Double click the RECORDINGS folder.

PHRASELATOR OPERATOR MAINTENANCE

11-47. The Phraselator has been built to withstand harsher wear and tear than most electronic devices; however, care must still be taken to minimize potential damage. The Phraselator has survived drop tests up to six (6) feet on concrete. Its components are shock-mounted to survive accidental drops and real-world abuse; however, do not deliberately drop or throw the device onto a hard surface.

11-48. Do not press sharp objects against the screen. The device is water-resistant, not waterproof. It can be cleaned or disinfected, used in a heavy rainstorm, or survive being dropped in a puddle. Do not immerse in water and do not place heavy objects on the Phraselator.

TROUBLE SHOOTING

11-49. To adjust the Stylus Tap Setting, follow the procedures described below:

- From within the Phraselator P2 v3.0 application, go to the OPTIONS tab. Tap the EXIT PHRASELATOR button and then the YES button.
- Tap the Windows icon at the bottom left of your screen.
- Tap SETTINGS and then the CONTROL PANEL option. A window with several icons will appear.
- Scroll down to find the STYLUS icon and double tap it.

- When the STYLUS PROPERTIES window appears, tap the DOUBLE TAP tab if it is not already visible and follow the instructions on the screen.
- If necessary, also tap the CALIBRATION tab and recalibrate the device using the instructions provided on the screen.
- To exit the STYLUS PROPERTIES window and save changes, tap the OK button in the upper right-hand corner of the screen. To exit without saving changes, tap the X in the upper right-hand corner.
- Continue to tap the X in the upper right-hand corner of each window that appears until you return to the P2 desktop.
- To return to the v3.0 application, double tap the icon on the upper left of the screen that looks like a Phraselator P2.

11-50. Tips for speaking phrases or verbal commands into the P2 are described below:

- Hold the P2 four (4) to six (6) inches from your mouth, positioned in front of you so that you can read the screen.
- Speak clearly.
- Pronounce words correctly.
- Push the PUSH TO TALK™ button before speaking the phrase.
- Speak in your normal voice. Speaking too loudly or softly can affect accuracy.
- Complete the entire phrase before releasing the PUSH TO TALK™ button.
- Allow the P2 to find and speak the complete translated phrase before you speak another English phrase. Pause until the status at the bottom left of the screen displays "Ready," and then speak another phrase.

Note: To stop the Phraselator when it is speaking a phrase, press the PUSH TO TRANSLATE™ button.

- Speak only phrases under the PHRASES tab. Do not omit or change words in the phrase.
- Ensure that the battery does not get too low. Extremely low battery power can affect the Phraselator's recognition of your voice.

11-51. Remember to speak phrases that are included in the category selected. The P2 will not recognize phrases that are not in the selected category.

DEMONSTRATE PHRASELATOR OPERATION

11-52. Once the Phraselator lesson is complete, the instructor demonstrates Phraselator operation by following these procedures:

- Turn Phraselator on.
- Select someone with a known second language. If no one is available, select Arabic.
- Instructor plays three (3) to five (5) phrases in the known second language (or Arabic).
- Have student verify what the phrases conveyed. Omit if using Arabic.

VOICE RESPONSE TRANSLATOR SPECIFICATIONS

11-53. The VRT is a speaker-dependent, one-way translator designed to assist Law Enforcement Officers in communicating with non-English speaking individuals. The VRT uses voice recognition technologies developed in the former Soviet Union.

11-54. The device, on initial operation, requires users to "program" or "train" the unit to their voice patterns for specific "trigger" phrases. When spoken, the VRT will respond to the trigger phrase with an audio (recorded human voice) translation in a complete command or sentence, in the selected language. For example, if the user is working in Spanish and says "registration" as the trigger phrase, the device's response would be "¿Puedo ver la registracion del vehiculo?" (May I see the vehicle registration?).

11-55. Because the device uses voice recognition, success in the field is highly dependent on the user saying the trigger phrases with the same inflection and volume as recorded at the time of programming.

DESCRIPTION

11-56. Features and characteristics of the VRT are described below:

- Reliable speech recognition in ambient noise over 100dB.
- Capacity to hold 125 languages and up to 1,000 phrases per language.
- System weight is 11 ounces.
- Oven tested at 175 degrees Fahrenheit for 24 hours while operating with no failure or performance degradation.
- Tested at negative 20 degrees Fahrenheit for 24 hours with no performance degradation.
- Laboratory and field-tested to ensure no electromagnetic interference problems with other equipment.
- Highly resistant to most environmental conditions, including heat, cold, sand and dust, saltwater, shock, and electrostatic discharge.

VRT EMPLOYMENT CONSIDERATIONS

11-57. The VRT is used to:

- Ask yes-no questions where the response can be given with gestures. For example, "Are you injured?" "Does he have a weapon?" or "Is the vehicle a car?"
- Give orders or instructions. For example, "Show me your hands."
- Ask questions when the response can be to point to something. For example, "Show me what caused the injury."
- Convey information. For example, "Keep moving."

OPERATE VRT

11-58. Initial commands include the following:

- Turn the VRT power on using the top switch.
- Press the red button up to eight times to select your user identification number.
- VRT will say, "INITIAL TRAINING, TO TRAIN PLEASE SAY THE FOLLOWING WORDS." (See command cards for a complete list of all 24 commands to be recorded).

Note: "Goodbye" is one of 12 initial commands to be recorded.

- Next, the VRT will say, "TO VERIFY, PLEASE SAY THE FOLLOWING WORDS."
- Repeat each word until the VRT says, "TRAINING COMPLETE."

11-59. Do not rush. Wait until the red light has gone out before repeating each command. If the VRT does not respond, repeat the command. If there is no response for 20 or 30 seconds, turn the unit off and begin again. The first recordings require the most repeats and volume. Stand up and use your diaphragm!

Event Commands

11-60. Use the procedures described below to record events commands:

- Record ALL of the initial commands, and then briefly press the red button to put the VRT into ON STANDBY mode. If you don't say anything, it will automatically go to STANDBY mode.
- Say, "Begin training." VRT says, "WHICH EVENT?"
- Say, "Medical." VRT says, "MEDICAL, TO TRAIN PLEASE SAY THE FOLLOWING WORDS" (see complete list of medical commands on the Medical Command List).
- Continue until all medical commands are recorded.
- VRT says, "TO VERIFY PLEASE SAY THE FOLLOWING WORDS." Repeat each word until the VRT says, "TRAINING COMPLETE."
- Repeat the above steps for each event on the Command Card.

11-61. Watch the LIGHT. When you hear the recording prompt, wait until the red light goes off before reciting the command. This allows the VRT to switch from play to record. Though this takes only a fraction of a second, do not begin speaking too quickly or your command may not be completely recorded.

ACTIVATING THE VRT SYSTEM

11-62. Turn on the VRT and try a few of the initial Commands that work at all times (see Initial Command Card). Follow these procedures:

- Say, "Different Language." VRT will say, "DIFFERENT LANGUAGES?"
- Say, "Creole." VRT will say, "CREOLE."
- Say, "Begin Directions." VRT will say in the selected language, "I AM SPEAKING THROUGH A DEVICE THAT TRANSLATES SELECTED PHRASES INTO (language being used). PLEASE RESPOND WITH HAND SIGNALS OR BY WRITING ANSWERS FOR ME. PLEASE NOD YOUR HEAD FOR YES AND SHAKE YOUR HEAD FOR NO."

Note: You will use the command "Begin Directions" every time you use the device. "Begin Directions" tells the person being addressed how the VRT will help you communicate with him/her.

11-63. Say, "Event." VRT will say, "WHICH EVENT?"

- Say selected situation, for example "Medical." VRT will say, "MEDICAL."
- Proceed to give medical commands listed on the Command Card.
- At will, you may say, "Different Language" or "Change Event."

Playing Emergency Phrase

11-64. The VRT has an emergency phrase that can be played continuously by any person. The emergency phrase is "Stay away from the Soldiers. If you advance any closer, we will be forced to use deadly force."

11-65. To start the emergency phrase, turn the translator on and instead of entering a User Number, hold the button down for at least five seconds. Let go of the button and the emergency phrase will play continuously. If using with a megaphone, you must press the megaphone button.

11-66. To turn off the emergency phrase, press the button again.

Troubleshooting

11-67. If the VRT does not recognize commands. The VRT is designed to work in an environment of approximately 100 decibels of background noise. If the area you are in requires hearing protection, then background noise is most likely 100 decibels or louder. Try moving to a less noisy area.

11-68. If set up was conducted in a quiet environment, then the VRT may not recognize commands. Set up again in an environment similar in noise to the one you will be operating in.

11-69. If set up was conducted with a calm voice, the VRT may not recognize an excited or loud command. Set up the VRT using the tone and level of voice being used in the operational environment.

11-70. Unit does not reset after button is held down for five or more seconds. Unit is not set on Standby or the button is not being fully pressed down. After pushing the button down for five or more seconds, the VRT will say "On standby." Press the button again for five or more seconds. When you release the button, the unit should say, "Initial training."

11-71. If it takes longer for the VRT to boot up or it seems to be recognizing too slowly or not at all, this is usually caused by the battery being run down. Charge the unit.

DEMONSTRATE VRT OPERATION

11-72. Once the Operate VRT lesson is complete, the instructor demonstrates VRT operation by following these procedures:

- Set the VRT to the instructor's voice.
- Turn translator on.
- Say, "Different language." VRT will say, "DIFFERENT LANGUAGES?"
- Say, "Spanish." VRT will say, "SPANISH."
- Say the following phrases:
  - "You injured?" VRT will say in Spanish, "ARE YOU INJURED?"
  - "Break up now." VRT will say in Spanish, "FAILURE TO BREAK UP NOW WILL RESULT IN YOUR ARREST."
  - "Show hands." VRT will say in Spanish, "SHOW ME YOUR HANDS."

11-73. If there are Spanish speaking students or AIs, have them verify that the VRT is translating correctly.

MAGNETIC AUDIO DEVICE SPECIFICATIONS

11-74. The MADLT-PMS1B is an active speaker with built-in 300S Power Amplifier and custom designed 3-channel professional Pre-amplifier/Mixer. The Control Panel can accept Dynamic or Condenser Microphone as well as two other Line level devices such as MP3 player, IPod, CD player, Phraselator, or VRT. Figure 11-2 provides a picture of the MAD, highlighting some of its features.

Chapter 11

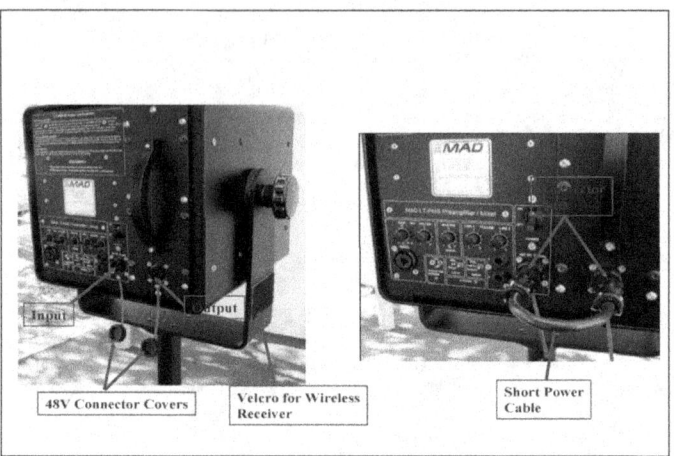

Figure 11-2. Features of the Magnetic Audio Device

MAD EMPLOYMENT CONSIDERATIONS

11-75. The MAD has various applications for both military and emergency response operations. The list below highlights several employment considerations:

- Support to military operations.
  - Internment and resettlement operations.
  - Checkpoint/roadblock operations.
  - Psychological operations.
  - Civil Affairs operations.
  - Border control operations.
  - Information dissemination.
  - Cordon and Search operations.
- Law Enforcement.
  - Special Reaction Team operations.
  - Crowd control.
  - Hostage negotiations/operations.
  - Public address system (ground or air).
  - Emergency evacuation.
- Emergency Services.
  - Unified command and control.
  - Victim search and rescue operations.
  - Civil defense public notification operations.

- Training.
  - Airborne drop zone operations.
  - Unimproved ranges and remote training sites.
  - Directing and instructing large groups of students.

OPERATE MAD

11-76. This section describes procedures for set up and employment of the MADLT-PMS1B. The preferred method of use is with the supplied tripod. Set up procedures include the following:

- Place tripod on stable surface. Stretch tripod's legs fully for better stability.
- Position speaker on tripod.
- Adjust proper height using retaining ring. (Horizontal aiming is easily done by simply rotating the speaker left or right.)

Note: To adjust vertical aiming angle, slightly loosen the Yoke Knobs and turn the Speaker up or down toward the target. When adjustment is done, tighten the Yoke Knobs.

The MADLT-PMS1B Speaker comes with installed, fully charged 48V Battery Pack and is ready for immediate use.

- Twist 48V Connector Covers to the left to expose the connectors. (If internally installed battery is empty or does not hold the charge, external MADLT-PMS1B 48V Battery Pack can be used instead.)
- Connect Battery output to the Speaker Power input with short power cable provided with the Unit. (Connectors are keyed and there is only one way to plug them properly. Twist the cable nuts clockwise until they click in place.)

Note: Before turning the power switch/breaker on, ensure that all potentiometers are turned all the way down.

- Flip the switch lever up. When the power is established, the Power LED will light green. (If the lever does not stay in the upper position, repeat step six (6) a couple of times until it catches on. The switch is a magnetic breaker and serves as a precise resettable fuse. During the startup time, the built-in amplifier creates a surge current that may trip the breaker.)
- Position microphone on head with capsule toward mouth. (For better performance and noise reduction, install the foam wind suppressor supplied with the microphone.)
- Plug microphone into MIC input.
  - Talk as loud as you would during voice broadcasting.
  - Turn MIC gain knob up until you see SIG/OVL LED flickering green.
  - Turn MIC volume knob to the 3 o'clock position.
  - Turn LOCK switch into MIC position.
  - Turn up Master Volume until desired loudness is reached.

Chapter 11

> **Note:** If feedback occurs, reduce master volume until feedback disappears. If SIG/OVL LED starts flickering green, don't increase the volume further. If this LED lights red, reduce master volume down until light disappears.

- For MP3 or other Line level devices, use Line inputs 1 or 2.
  - Turn LOCK switch to the Off position. Reduce Master volume to zero. Turn Line 1 volume knob to the 3 o'clock position.
  - Start MP3 player and adjust Master volume to desired loudness. If the sound is not loud enough turn Line 1 volume to maximum. For even louder output, increase MP3 output level.
- Turn on the siren.
  - Reduce the Master volume to zero first.
  - Turn the LOCK switch to Siren position.
  - Select one of the three (3) available tones with Siren switch.
  - Increase Master volume to desired level.
- To mix Microphone and MP3 signals simultaneously:
  - Reduce Master volume to low level.
  - Turn LOCK switch to MIC position.
  - Turn MIC volume and Line 1 volume knobs to mid-position.
  - Talk through microphone and increase microphone volume if needed. At the same time, adjust Line 1 volume to get proper loudness compared to voice output.

11-77. Figure 11-3 depicts the MAD set up procedures.

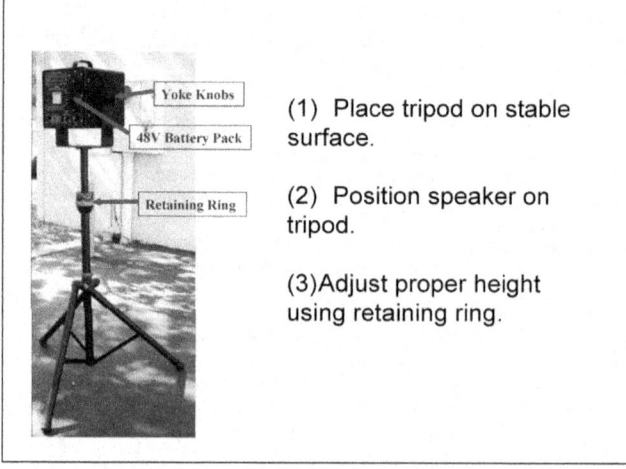

(1) Place tripod on stable surface.

(2) Position speaker on tripod.

(3) Adjust proper height using retaining ring.

Figure 11-3. MAD Set Up Procedures

Wireless Receiver

11-78. Mount the Wireless Receiver on Velcro Pad on the Yoke side. Position receiver so the antenna points toward target. Connect receiver's output to MIC input on the speaker with supplied cable (4 pin Hirose Connector on the receiver side and XLR connector on the speaker side). Adjust receiver's output to 3 o'clock.

11-79. To monitor the signal transmission, plug MP3's headphones into Receiver's Monitor Output. Adjust Monitor level to comfortable loudness. Receiver uses 4 x AA batteries. External 12V DC power can be used if AC power is available.

11-80. Calibrate Speaker MIC Gain the same way as with wired microphone. Head worn Microphone with 4 pin Hirose connector needs to be plugged into transmitter. Both Receiver and Transmitter need to be turned on and they must have the same carrier frequency. Transmitter needs to be set for MIC input as a source. MIC input is default setting for supplied Wireless Transmitter.

11-81. Wireless Transmitter can be used to transmit sound from MP3 player. A special cable is provided for this purpose.

DEMONSTRATE MAD OPERATION

11-82. Once the MAD lesson is complete, the instructor demonstrates placing MAD into operation. Set up and turn on MAD following these procedures:

Individual at 200 Meters:

- Place AI at 200 meters.
- Give simple commands to AI without MAD.
- "Hands on top of your head."
- "Kneel down, facing me."
- Repeat commands using MAD.

Individual in Vehicle at 200 Meters:

- Place AI in vehicle 200 meters from MAD.
- Give AI commands:
  - "Turn on headlights."
  - "Flash lights twice."
  - "Honk horn once."

11-83. Once the instructor completes this demonstration. Operate MADLT-PMS1B with Phraselator to combine both capabilities. Follow previous guidance on Phraselator using MAD to amplify the phrases.

This page intentionally left blank.

Appendix A

Nonlethal Capabilities Set

NONLETHAL CAPABILITIES SET MODULES

A-1. Non-lethal capabilities are available for use in a variety of conflict scenarios, from humanitarian and peace operations to major combat operations. Service-unique NLCS support platoon-company size units and contain a mix of counter-personnel and counter-materiel systems, protective equipment, enhancement devices, and training devices. Some of the NLCS counter-personnel items include counter materiel devices, including tire spikes (caltrops) and the portable vehicle-arresting barrier, both of which are used to deny vehicles access to critical infrastructure at roadblocks and access control points.

A-2. Currently, there are initiatives to develop and package mission-specific sets. The available systems range from pre-emplaced munitions to vehicle arresting devices. NLWs are multi-capable, with the ability to strike a single target or multiple targets.

A-3. One (1) NLCS consists of four (4) distinct mission modules and two (2) counter-personnel sub-modules.

A-4. The mission modules include:

- Four (4) Checkpoint Operations Modules each consisting of four (4) Hasty Checkpoint Sub-Modules and one (1) Deliberate Checkpoint Sub-Module.
- Four (4) Mounted (convoy operations) Modules.
- Two (2) Crowd Control/Detainee Operations Modules.
- Four (4) Dismounted Operations Modules.

CHECKPOINT OPERATIONS (DELIBERATE)

A-5. The following components are included in the Deliberate Checkpoint Operations Sub-Module:

- One (1) Entry Point Vehicle Kit.
- Four (4) Orange Safety Vests.
- 30 Orange Safety Cones.
- Three (3) Handheld Metal Detectors.
- Two (2) Magnum Spike Systems (MS-16).
- One (1) Alternating Current (AC) Generator.
- Six (6) Portable Light Sets.
- One (1) Phraselator Model 2000.
- One (1) Voice Response Translator.
- One (1) Magnetic Audio Device.
- 200 Caltrops.
- 10 LyteFlares.
- One (1) High Intensity Light.

Appendix A

- Two (2) Traffic Paddles (Lighted).
- One (1) M1 Portable Vehicle Arresting Barrier.

CHECKPOINT OPERATIONS (HASTY)

A-6. The following components are included in the Hasty Checkpoint Operations Sub-Module:

- One (1) Expeditionary Vehicle Inspection Toolkit.
- Four (4) Orange Safety Vests.
- Three (3) Hand Held Metal Detectors.
- Two (2) Magnum Spike Systems (MS-16).
- One (1) Phraselator Model 2000.
- One (1) Voice Response Translator.
- One (1) Magnetic Audio Device.
- One (1) M2 Vehicle Lightweight Arresting Device.
- 200 Caltrops.
- 10 LyteFlares.
- One (1) High Intensity Light.
- Two (2) Traffic Paddles (Lighted).
- Two (2) Vehicle Inspection Checklists.

MOUNTED OPERATIONS

A-7. The following components are included in the Mounted (Convoy) Operations Module:

- 10 Magnetic Audio Devices.
- 10 Go Lights.

DISMOUNTED OPERATIONS

A-8. The following components are included in the Dismounted (Patrol) Operations Module:

- One (1) High Intensity Light.
- One (1) Phraselator Model 2000.
- One (1) Voice Response Translator.
- 1,000 Disposable Restraints.
- 10 Disposable Restraint Cutters.
- 10 Nonlethal Munitions Pouches.

CROWD CONTROL AND DETAINEE OPERATIONS

A-9. The following components are included in the Crowd Control/Detainee Operations Module:

- 30 Non-Ballistic Riot Face Shields. Platoon size.
- 30 Non-Ballistic Riot Body Shields.
- 30 Non-Ballistic Riot Shin Guards.
- 30 Expandable Riot Batons.
- One (1) Handheld Bullhorn.
- One (1) Magnetic Audio Device.
- One (1) Phraselator Model 200.
- One (1) Voice Response Translator.
- 30 Mk-4 Pouches (for M39 Individual Riot Control Agent Disperser.
- One (1) Hobble (leg restraint).
- One (1) Full Body Cuff (restraint).
- 2,000 Disposable Restraints.
- 10 Disposable Restraint Cutters.
- 10 Nonlethal Munitions Pouches.

ELECTRODE STUN DEVICE

A-10. The following components are included in the Electrode Stun Device Sub-Module:

- Six (6) X26 E TASERS®.
- Six (6) XD Power Modules.
- Six (6) Holsters.
- One (1) Data Download Unit.
- 150 25' XP Cartridges.
- 150 21' Standard Cartridges.

ARMORER'S KIT

A-11. The following components are included in the Armorer's Kit:

- One (1) AA Lithium Battery Four-Pack.
- Three (3) Metal Detector End Caps.
- One (1) Phraselator Battery Door.
- Two (2) Phraselator Styli.
- One (1) Adjustable Wrench.
- Two (2) Phillips Screwdrivers.
- One (1) Flat Screwdriver.

- One (1) Owner's Manual.
- One (1) Pelican Case.

ACCOUNTABILITY

A-12. Accountability is a reoccurring issue with many high-dollar expendable items. Many of the items found in the NLCS are high dollar, expendable items that can be resupplied through standard supply channels or General Services Administration purchase.

Appendix B
Nonlethal Munitions Practical Exercise

NONLETHAL MUNITIONS TRAINING AND FAMILIARIZATION

B-1. Placing single and group targets in 10-meter increments from 10 meters out to 50 meters is the suggested training strategy.

B-2. Inform the students that because they will be firing different rounds with different range capabilities, particular attention must be paid to the rounds they are loading into their weapons while conducting the live fire exercise.

B-3. Inform the students that they will be required to place five (5) each M1006 Sponge Grenades, M1012 12-gauge point rounds and two (2) each of the M1029 Crowd Dispersal Cartridges, M1013 Crowd Dispersal Cartridge rounds on the designated targets within their firing lane.

B-4. Instructor follows the procedures described below:

- Inform the students that they will be required to engage targets with various nonlethal munitions.
- Perform a safety briefing prior to the start of the practical exercise:
  - "At any time during this practical exercise, if you do not understand a step, raise your non-firing hand for assistance, and an instructor will provide help."
- Inform the students that they will have four (4) hours to complete the practical exercise.
- Allow 10 minutes at the end of the practical exercise to review, answer student questions, and correct student misunderstandings.
- Critique and assist students as necessary throughout the practical exercise.
- The person in charge must clear the training area before each order.

INSTRUCTOR FIRING COMMANDS

B-5. The narrations below provides the commands and procedures for conducting the exercise:

- "Firers, you should be standing directly behind your firing point number, facing the targets."
- "Firers, you will fire the 40mm Grenade Launcher from the standing unsupported firing position. Secure your weapon." (Pause and observe firing line).
- "Open the action of your weapon and load your weapon." (Pause and observe).
- "Firers, on the command of "Targets!" you will do the following:"
  - "Bring your weapon from a low ready to firing position."
  - "Disengage the safety."
  - "You will fire five (5) M1006 Sponge Grenade rounds. You will fire two (2) rounds at your 10-meter target, two (2) rounds at your 20-meter target and one (1) round at your 50-meter target, reloading and firing as necessary."
  - "You will reload and fire two (2) M1029 Crowd Dispersal Cartridge rounds. You will fire two (2) rounds at your 10-meter target."

Appendix B

- "Firers, after you have completed firing, place your weapon on SAFE, raise your non-firing hand, and wait to be cleared by a safety. After you have been cleared by a safety, place your weapon on the ground with the muzzle pointed downrange then turn and face the tower."

Engage Targets with 40mm Nonlethal Munitions

- "Are we ready on the right?" Instructor will wait for signal from the right side of the firing line.
    - If the right is ready, say, "The right is ready."
    - If not, say, "The right is not ready." Wait until the right is ready, as signaled by a safety assistant, then say, "The right is ready."
- "Are we ready in the center?" Instructor will wait for signal from the center of the firing line.
    - If the center is ready, say, "The center is ready."
    - If not, say, "The center is not ready." Wait until the center is ready, as signaled by a safety assistant, then say, "The center is ready."
- "Are we ready on the left?" Instructor will wait for signal from the left side of the firing line.
    - If the left is ready, say, "The left is ready."
    - If not, say, "The left is not ready." Wait until the left is ready, as signaled by a safety assistant, then say, "The left is ready."
- "The firing line is ready."
- Have firers secure and load their weapons. Give the command, "Targets!" (Allow time for firing).
- Give the command, "Cease Fire!"
- "Firers, if you have rounds remaining, place the weapon on SAFE and raise your non-firing hand. You will get a chance to fire the remaining rounds only if your weapon malfunctioned.
- "Safeties, move forward and check all alibis." (Pause and observe).
- "Alibi firers, on the command of "Targets!" you will have 30 seconds to fire these rounds."
- Give the command, "Targets!" (Allow 30 seconds for firing).
- Give the command, "Cease Fire!"
- "Firers, place your weapon on SAFE, raise your non-firing hand, and wait to be cleared by a safety. After you have been cleared by a safety, place your weapon on the ground with the muzzle pointed downrange then turn and face the tower."
- "Firers, once you have been cleared by a safety, place your weapon on the ground with the muzzle pointed downrange then turn and face the tower."
- "Are we clear on the right?" Instructor will wait for signal from the right side of the firing line.
    - If the right is clear, say, "The right is clear."
    - If not, say, "The right is not clear." Wait until the right is clear, as signaled by a safety assistant, then say, "The right is clear."
- "Are we clear in the center?" Instructor will wait for signal from the center of the firing line.
    - If the center is clear, say, "The center is ready."
    - If not, say, "The center is not clear." Wait until the center is clear, as signaled by a safety assistant, then say, "The center is clear."

- "Are we clear on the left?" Instructor will wait for signal from the left side of the firing line.
  - If the left is clear, say, "The left is clear."
  - If not, say, "The left is not clear." Wait until the left is clear, as signaled by a safety assistant, then say, "The left is clear."
- "The firing line is clear."
- "Firers, walk to the center of the firing line, then move toward the assembly area and wait for further instructions."
- "At any time during this practical exercise, if you feel lost or do not understand a step, raise your non-firing hand for assistance, and an instructor will provide help."
- "Firers, you should be standing directly behind your firing point number, facing the targets."

Engage Targets with 12-Gauge Shotgun

- "Firers, you will fire the 12-gauge shotgun from the standing unsupported firing position. Secure your weapon." (Pause and observe firing line).
- "Open the action of your weapon and load your weapon." (Pause and observe firing line).
- "Firers, on the command of "Targets!" you will do the following:"
  - "Bring your weapon from a low ready to firing position."
  - "Disengage the safety."
  - "You will fire five (5) M1012 12-gauge point rounds. You will fire two (2) rounds at your 10-meter target, two (2) rounds at your 15-meter target, and one (1) round at your 20-meter target. Reload and charge the weapon as necessary."
  - "You will reload and fire two (2) M1013 Crowd Dispersal Cartridge rounds. You will fire two (2) rounds at your 10-meter target."
- "Firers, after you have completed firing, place your weapon on SAFE, raise your non-firing hand, and wait to be cleared by a safety. After you have been cleared by a safety, place your weapon on the ground with the muzzle pointed downrange, then turn and face the tower."
- "Are we ready on the right?" Instructor will wait for signal from the right side of the firing line.
  - If the right is ready, say, "The right is ready."
  - If not, say, "The right is not ready." Wait until the right is ready, as signaled by a safety assistant, then say, "The right is ready."
- "Are we ready in the center?" Instructor will wait for signal from the center of the firing line.
  - If the center is ready, say, "The center is ready."
  - If not, say, "The center is not ready." Wait until the center is ready, as signaled by a safety assistant, then say, "The center is ready."
- "Are we ready on the left?" Instructor will wait for signal from the left side of the firing line.
  - If the left is ready, say, "The left is ready."
  - If not, say, "The left is not ready." Wait until the left is ready, as signaled by a safety assistant, then say, "The left is ready."
- "The firing line is ready."

Appendix B

- Have firers secure and load their weapons. Give the command, "Targets!" Instructor will allow time for firing.
- Give the command, "Cease Fire!"
- "Firers, if you have rounds remaining, place the weapon on SAFE and raise your non-firing hand. You will get a chance to fire the remaining rounds only if your weapon malfunctioned.
- "Safeties, move forward and check all alibis." (Pause and observe firing line).
- "Alibi firers, on the command of "Targets!" you will have 30 seconds to fire these rounds."
- Give the command, "Targets!" Instructor will allow 30 seconds for firing.
- Give the command, "Cease Fire!"
- "Firers, place your weapon on SAFE, raise your non-firing hand, and wait to be cleared by a safety. After you have been cleared by a safety, place your weapon on the ground with the muzzle pointed downrange then turn and face the tower."
- "Firers, once you have been cleared by a safety, place your weapon on the ground with the muzzle pointed downrange then turn and face the tower."
- "Are we clear on the right?" Instructor will wait for signal from the right side of the firing line.
  - If the right is clear, say, "The right is clear."
  - If not, say, "The right is not clear." Wait until the right is clear, as signaled by a safety assistant, then say, "The right is clear."
- "Are we clear in the center?" Instructor will wait for signal from the center of the firing line.
  - If the center is clear, say, "The center is clear."
  - If not, say, "The center is not clear." Wait until the center is clear, as signaled by a safety assistant, then say, "The center is clear."
- "Are we clear on the left?" Instructor will wait for signal from the left.
  - If the left is clear, say, "The left is clear."
  - If not, say, "The left is not clear." Wait until the left is clear, as signaled by a safety assistant, then say, "The left is clear."
- "The firing line is clear."
- "Firers, walk to the center of the firing line, then move toward the assembly area and wait for further instructions."

NONLETHAL MUNITIONS QUALIFICATION RANGE CHECKLIST

B-6. Table B-1 provides a nonlethal munitions qualification range checklist.

Table B-1. Nonlethal Munitions Qualification Range Checklist

Nonlethal Munitions Qualification Range	
25 Meter Qualification Range	
M203 (1 per 5 Soldiers)	
12 gauge Shotgun (1 per 5 Soldiers)	
24 E-type Silhouettes with stands and sandbags (4 per lane)	
Combat Lifesaver (with bag)	
Trash Bags	
Range Detail	
Range Safeties (1 per lane)	
NOTE: These are minimum Requirements.	

Appendix B

QUALIFICATION SCORE CARD

B-7. Table B-2 illustrates a typical nonlethal qualification score card.

Table B-2. Nonlethal Qualification Score Card

NAME (Last, First, MI)				DATE	
LANE NO.	FIRING ORDER		UNIT	SSN	
TABLE 1 – M1006 Point Round					
		30 Meter Target	1st Attempt Hits 2nd Attempt Hits 3rd Attempt Hits		
		20 Meter Target	1st Attempt Hits 2nd Attempt Hits 3rd Attempt Hits		
		10 Meter Target	1st Attempt Hits 2nd Attempt Hits 3rd Attempt Hits		
NOTE: In order to receive a Go on the M1006 Table, Soldiers must hit 4 out of 5 targets.			1st Attempt GO_____	NO-GO_____	
			2nd Attempt GO_____	NO-GO_____	
			3rd Attempt GO_____	NO-GO_____	
TABLE 1 – M1012 Point Round					
		20 Meter Target	1st Attempt Hits 2nd Attempt Hits 3rd Attempt Hits		
		15 Meter Target	1st Attempt Hits 2nd Attempt Hits 3rd Attempt Hits		
		10 Meter Target	1st Attempt Hits 2nd Attempt Hits 3rd Attempt Hits		
NOTE: In order to receive a Go on the M1012 Table, Soldiers must hit 4 out of 5 targets.			1st Attempt GO_____	NO-GO_____	
			2nd Attempt GO_____	NO-GO_____	
			3rd Attempt GO_____	NO-GO_____	
Firer's Signature			Scorer's Signature		

Appendix C
M5 MCCM Practical Exercise

PRACTICAL EXERCISE

C-1. M5 certification is an annual training requirement. This exercise requires an area large enough for five (5) training lanes. Students will need to take cover no less than 50 feet (15 meters) behind the M5 MCCM. Instructor follows the procedures described below:

- Inform the students that they will be required to employ the M5 MCCM.
- Conduct a safety briefing prior to the start of the practical exercise.
- Inform the students that they will have 20 minutes each to complete the practical exercise.
- Allow 10 minutes at the end of the practical exercise to review, answer student questions, and correct student misunderstandings.
- Critique and assist students as necessary throughout the practical exercise.

SAFETY

C-2. Perform a safety briefing prior to the start of the practical exercise. Inform the students:

- "At any time during this Practical Exercise, if you do not understand a step, raise your hand for assistance, and an instructor will provide help."
- "Striking, yanking, or abusing the blasting cap may cause detonation."
- "Lethal trauma may occur to individuals within five (5) meters of the front of the munition when fired."
- "Keep hands clear of the shock tube insertion point when detonating the munition; there is a possibility of hot, gaseous by-products being released."

EMPLACE, AIM, ARM, AND FIRE THE M5 MCCM

INSTRUCTOR FIRING COMMANDS

C-3. The following narrations provide the commands and procedures for conducting the exercise.

Install M5 MCCM

- "Ensure that you have one (1) M18A1 practice claymore mine, a shock-tube/blasting cap assembly, an M81 igniter, a shock tube cutter, a supplemental instruction sheet, and a shock tube sealing nut (contained in a barrier bag). Repack all items in your bandoleer and place on your shoulder."
- "Firers, ensuring that you have your bandoleer, move forward pacing off 15 feet (five (5) Meters)."
- "Pick up an object, such as a rock or stick, and place on the ground to mark the position. This is the minimum engagement line for the M5 MCCM."
- "Firers, move back to your original position and face toward the threat."
- "Remove the M5 MCCM from the bandoleer and the barrier bag."

Appendix C

- "Open the M5 MCCM's legs and separate them at the hinged points, ensuring they are open to their widest point."
- "Set the M5 MCCM in position, and push the legs firmly into the ground, maintaining a fist height from the ground to the bottom of the M5 MCCM."
- "Roughly aim the M5 MCCM."
- "Remove one of the shipping plug's priming adapters, either side will work."
- "Place the blasting cap in the adapter and slip the shock tube into the groove."
- "Thread the adapter back into the cap well of the M5 MCCM."

Aim M5 MCCM

- "Firers, position your eye about 15 to 23 centimeters to the rear of the M5 MCCM sight."
- "Aim the munition by aligning the front and rear edges of the sight with the aim point."
- "Align the front and rear sights to aim."
- "Select an aiming point about 15 feet (5 meters) in front of the munition and about 30 centimeters above the ground or boot/calf height."
- "Place a sandbag 30 centimeters behind the M5 MCCM."
- "Place a rock, a double handful of dirt, or a sandbag on the shock tube, about 60 centimeters from the M5 MCCM, to keep it stationary."
- "Continue to unroll the shock tube assembly back to a safe detonating distance, 50 feet (15 meters) if behind a barrier or 100 feet (30 meters) if no barrier is used."
- "Place the empty spool into the bandoleer pocket for reuse if the munition is recovered."

> **"WARNING: Damaging the shock-tube when securing it with a heavy object may cause it to malfunction."**

Arm M5 MCCM

- "Remove the barrier bag containing the M81 igniter, the supplemental instruction sheet, the cutter, and the sealing nut from the bandoleer."
- "Cut the barrier bag one (1) centimeter from the seal so that the bag can be reused if the shock tube is recovered. Leave the sealing nut in the bag in case the shock tube is recovered."
- "Remove the M81 igniter, the supplemental instruction sheet, and the shock tube cutter from the barrier bag."
- "Loosen, but do not remove, the M81 igniter fuse cap holder (soft clear plug) by turning it counter-clockwise three (3) to four (4) full turns."
- "Pull, do not twist, and remove the plastic weatherproofing plug (solid green plug) from the fuse cap holder. Do not remove the fuse cap holder. Place the plug in the barrier bag for possible reuse."
- "Place the barrier bag back into the bandoleer pocket."
- "Make a straight, 90 degree cut about 15 to 20 centimeters off the crimped/sealed end of the shock tube. Do not cut the tube at an angle because the M81 igniter may not ignite the shock tube."

M5 MCCM Practical Exercise

- "Insert the open end of the shock tube all the way into the hole of the M81 igniter until it stops, about three (3) centimeters. Twist while pushing the shock tube in to ensure that it makes good contact with the M81 igniter primer."
- "Tighten the M81 igniter fuse cap holder (shipping plug adapter). The end of the shock tube must sit fully in the M81 igniter fuse cap holder to properly initiate the shock tube. Do not hold the shock tube when firing it."

Fire M5 MCCM

- "Remove the cotter pin safety from the igniter and store it in the barrier bag."
- "Take cover and yell, "FIRE IN THE HOLE!""
- "Push the plunger ring in then twist one-quarter-turn. Fire by pulling sharply on the M81 igniter."

Misfire Procedures

C-4. "Use following procedure if the M81 misfires:"

- "Re-cock the M81 igniter and attempt to fire again; the M81 igniter can be re-cocked twice. If it still does not fire and an additional M81 igniter is available, start misfire procedures."
- "Remove the original M81 igniter and measure about one (1) meter down the shock tube."
- "Make a straight, 90 degree cut at that point with the shock tube cutter. Cut an additional 15 centimeters of shock tube off and conduct a blow test."
- "Blow through those 15 centimeters of shock tube with the open end in the palm of your hand."
- "If you notice a small amount of silver powder, reinstall a new M81 igniter and attempt to fire. If you do not notice silver powder, repeat the one (1) meter measurement and cut and blow test that piece of the shock tube until some silver powder shows. This can be done two (2) additional times for a total of three (3) meters."
- "Install a new M81, and attempt to fire. If the M5 still does not function, wait 30 minutes before approaching, marking the M5, and notify the Explosive Ordnance Disposal (EOD) Office for proper disposal procedures."
- "If an additional M81 igniter is not available, wait 30 minutes before approaching, mark the M5 MCCM, and notify the EOD office for proper disposal procedures."

> **"WARNING: Ingesting or inhaling the explosive powder in the shock tube may cause illness. Do not breathe in when applying the blow test because explosive powder may be swallowed instead of being blown out through the tube. Wash hands before eating or drinking after handling the shock tube."**

Disarm and Recover

- "Disarm and recover the M5 MCCM by following the installation and arming steps in reverse order. Do not disassemble the M5 MCCM. Return all items to the bandoleer."

This page intentionally left blank.

Appendix D
X26E TASER Performance Evaluation

X26E TRAINING AND QUALIFICATION RECORD

D-1. This training is an annual training certification requirement. Set up four (4) stations or points in sequence for performance evaluation. Ensure that a properly trained and certified assistance instructor accompanies the firer through the course of fire until completion of the exercise.

D-2. The target material or backstop that you use must allow the probe barbs to reliably penetrate/stick. Test it before training. Appropriate eye protection is required for everyone—firers, coaches, safeties, instructors, and anyone who enters the immediate training vicinity (a safety zone that you establish and mark).

SAFETY

- This exercise does not require a live-fire small arms range. Course managers and instructors must ensure that all applicable safety requirements and qualifications/certification for safety personnel are met.
- Perform a safety briefing prior to the start of the practical exercise. Inform the students:
    - "At any time during this Practical Exercise, if you feel lost or do not understand a step, raise your non-firing hand for assistance, and an instructor will provide help."
- The person in charge must clear the training area before each order.

PERFORMANCE DRILLS

Station 1: Loading and Unloading Procedures – Soldiers are issued an expended TASER ® cartridge, and must demonstrate proper loading and unloading procedures ensuring that they grasp the top and bottom of the cartridge and do not place their hands in front of the cartridge.

Station 2: Proper Sight Picture – Soldier are issued one training TASER ® cartridge. Without using the laser guide, Solders pick up a good sight picture on a target at eight meters. Soldiers engage target. In order to receive a go at this station, both TASER ® probes must hit the torso portion of the target. No head shots.

Station 3: Misfire Procedures – Soldiers are issued one (1) expended cartridge and (1) one 15 meter cartridge. The Soldier loads the expended cartridge to simulate a misfire. Soldiers must unload the cartridge, reload the 15 meter cartridge and engage a target at 15 meters. The probes must impact the torso area of the target. No head shots.

Station 4: Multiple Targets – Soldiers are issued a 25 and 35 meter cartridge. They must engage multiple targets positioned at 15 and 20 meters. The probes should impact the torso of the target with no head shots.

RANGE CHECKLIST

D-3. Table D-1 provides the minimum resource requirements for TASER Qualification.

Table D-1. TASER Qualification Range Checklist

TASER QUALIFICATION CHECKLIST	
4 E-type Silhouettes with stands	
Combat Lifesaver (with bag)	
Alcohol Swabs	
Trash Bags	
Rubber Gloves	
TASERs® (at least 1 per 5 firers)	
TASER ® cartridges	
Practice Cartridges – 2 per firer	
25 Meter Cartridges – 1 per firer	
35 Meter Cartridges – 1 per firer	

Appendix E
Components of the Checkpoint Operations Mission Module

E-1. Each nonlethal Checkpoint Operations Mission Module provides the necessary components required for both deliberate and hasty checkpoint operations. During the introduction of training for the NLCS Checkpoint Operations Mission Module, Soldiers will familiarize themselves with and use all components of the Checkpoint Operations Mission Module in a simulated scenario at a hasty or deliberate checkpoint. These components and procedures apply to vehicle checkpoints, roadblocks, tactical control points, and other types of access control points.

E-2. Each NLCS includes four Checkpoint Operations Mission Modules. Each Module includes one (1) Deliberate Checkpoint Sub-Module and four (4) Hasty Checkpoint Sub-Modules.

DELIBERATE CHECKPOINT SUB-MODULE COMPONENTS

E-3. The Deliberate Checkpoint Sub-Module contains the following parts:

- Thirty (30) 28-inch Orange Safety Cones.
- One (1) AC Generator Set 120 VAC 60 H with grounding rod.
- One (1) Entry Point Vehicle Kit.
- Four (4) Orange Safety Vests.
- Three (3) Handheld Metal Detectors.
- Two (2) Magnum Spike Systems (MS-16).
- Six (6) Portable Light Sets.
- One (1) Phraselator Model 2000.
- One (1) Voice Response Translator (VRT).
- One (1) Magnetic Audio Device (MAD).
- Two Hundred (200) Caltrops.
- Ten (10) LyteFlares.
- One (1) High Intensity Light.
- Two (2) Traffic Paddles.
- One (1) M1 Portable Vehicle Arresting Barrier (PVAB).
- Four (4) Vehicle Inspection Guides.

HASTY CHECKPOINT SUB-MODULE COMPONENTS

E-4. Each of the four (4) Hasty Checkpoint Sub-Module components contains the following parts:

- One (1) Expeditionary Vehicle Inspection Toolkit.
- Sixteen (16) Orange Safety Vests.
- Twelve (12) Handheld Metal Detectors.
- Eight (8) Magnum Spike Systems (MS-16).
- Four (4) Phraselator Model 2000.

Appendix E

- Four (4) VRTs.
- Four (4) MADs.
- Four (4) M2 Vehicle Lightweight Arresting Device (VLAD).
- Eight hundred (800) Caltrops.
- Forty (40) LyteFlares.
- Four (4) High Intensity Light.
- Eight (8) Traffic Paddles.
- Eight (8) Vehicle Inspection Guides.

COMPONENT CHARACTERISTICS

E-5. **AC Generator.** The AC generator is a common 2-kilowatt Army generator that comes complete with all component parts including a grounding rod. A hammer is not included. Soldiers must carry a hammer or picket driver with their NLCS Checkpoint Operations equipment in order to drive the grounding rod into the ground.

E-6. Before operating the generator, Soldiers must consult with their local policy for licensing requirements. Any Soldier who will operate the generator must be licensed in accordance with regulations and local policies.

E-7. **High Intensity Lights.** There are three types of lights: handheld, vehicle-mounted, and tripod-mounted.

- Handheld flashlights are included in the Checkpoint Operations Mission Module. These flashlights:
  - Are used in periods of low light.
  - Can be used to illuminate dark areas and aid in the inspection of personnel, property, and vehicles.
  - Temporarily blind or limit the vision of a subject(s).
- Vehicle-mounted high intensity lights:
  - Illuminate area in periods of low light.
  - Temporarily blind or limit the vision of a subject(s).
  - Can be used as a mobile light source.
- Tripod-mounted high intensity lights are:
  - Used to illuminate area during periods of low light (floodlight).
  - Mainly employed at static positions.

E-8. **LyteFlares.** The LyteFlare is a palm-sized non-incendiary flare that does not produce noxious smoke or hot spark. They are highly visible, long lasting (over 400 hours on two AA batteries), and reusable. LyteFlares resist rain, snow, fuel contamination, and vehicles driving over them. They can be used in periods of low light and limited visibility to supplement traffic cones and other established traffic control devices and barriers.

E-9. **Vehicle Arresting Devices.** Vehicle Arresting Devices are rapidly and easily deployed and recovered. There are two types currently in use.

- Caltrops:
  - Caltrops can be hand-scattered and when employed, always land with the point up. They are "V" shape by design to ensure rapid tire deflation, incapacitating the target vehicle.

Components of the Checkpoint Operations Mission Module

- Caltrops are used for vehicle denial, and they supplement permanent or portable barriers or other arresting devices like the PVAB or VLAD. Several Caltrops strung together can be used alone at a hasty checkpoint.
- Training on Caltrops is included in the Identify Arresting Vehicle Devices Lesson Plan.

■ Magnum Spike Systems:

- Magnum Spike Systems are hand-rolled over a target area measuring 16-feet long in length. They can be used to supplement permanent or portable barriers and other arresting devices, such as the PVAB or VLAD, and are especially useful in establishing a hasty checkpoint.
- The device has a hollow-spike design, which allows rapid deflation. Each device contains 160 detachable spikes.
- Training on the Magnum Spike System is included in the Identify Arresting Vehicle Devices Lesson Plan.

Note: Caltrops and the Magnum Spike System are designed to flatten tires and may not stop all vehicles.

E-10. **Portable Vehicle Arresting Barrier.** The PVAB can be placed on any flat surface and in any environment. It is designed for vehicle denial and can successfully stop vehicles weighing up to 7,500 lbs. at speeds up to 45 mph within 200 feet. The standby mode of PVAB allows normal traffic to pass through regardless of volume without damaging the PVAB. The PVAB is used as a bi-directional system capable of one-lane (15 feet minimum) or two-lane (24 feet maximum) coverage and is remotely activated up to 300 ft. away. The PVAB is most useful at entry control points, deliberate checkpoints, and at the perimeter of controlled access areas. It can be anchored to alternate objects like guardrails or supports, bridge abutments, trees, or a parked vehicle such as a high mobility multipurpose wheeled vehicle. The PVAB can be placed in operation in less than one hour by two trained personnel. PVAB training is included in the Identify Arresting Vehicle Devices Lesson Plan.

E-11. **Vehicle Lightweight Arresting Device.** The VLAD is designed to stop vehicles less than two tons predominately in a straight line. The site selected should have 330 feet (approximately 100 meters) of straight road beyond the net device. In wet conditions, the stopping distance should be doubled. The VLAD vehicle arresting system can be effective at stopping the "majority" of passenger vehicles and light trucks and is particularly useful in supplementing the PVAB or used alone in a hasty checkpoint. The lanyard deployment mode of the VLAD allows all vehicle traffic to pass by regardless of volume without damaging the VLAD. VLAD training is included in the Identify Arresting Vehicle Devices Lesson Plan.

E-12. **Handheld Metal Detector.** Handheld metal detectors are used to rapidly scan personnel for metallic weapons and metallic components of explosive devices.

E-13. **Phraselator Model 2000.** The Phraselator device is capable of voice translation through voice response or pre-programmed phrases and incorporates "Closed ended" questions that can be easily answered with physical gestures to aid in conversations and on-the-spot interviews with indigenous, non-English speaking personnel. Various language modules are available for use with the Phraselator. The device weighs one pound and has a directional microphone to block out background noise. The Phraselator is capable of universal voice recognition regardless of accent. Phraselator training is included in the Identify Communication Devices Lesson Plan.

E-14. **Voice Response Translator.** The VRT is a portable electronic translation device that emits short, prerecorded phrases in one of 26 languages. The VRT is eyes- and hands-free when using the headset. It can be connected to a separate portable megaphone or the Acoustic Hailing Device for louder transmission; however, it is more frequently used as a stand-alone device. Each device accommodates up to eight (8) different users. It is programmable for up to 350 phrases per language and has a 95% translation accuracy. The VRT is *not* a direct translator or a substitute for a live interpreter. It uses preselected keywords to generate specific phrases and/or questions that prompt a yes or no answer. The device is like a talking phrase dictionary that translates

Appendix E

predetermined phrases from English into a specific target language. The VRT does *not* translate target languages into English.

E-15. **Magnetic Audio Device.** MAD is a directional audio device capable of transmitting verbal messages or preprogrammed audible tones and can be used to discern intent or provide verbal warnings at established standoff ranges. Messages and tones can be transmitted directly through the MAD unit or wirelessly with the included wireless transmitter headset. MAD accepts MP3, VRT, Phraselator, and other auxiliary devices. The MAD can be mounted on the included tripod or on a metal-skinned vehicle using the magnetic mounting system.

E-16. **Handheld Inspection Lights.** Various light systems can be worn or used by the Soldier to aid in routine vehicle inspections and in inspecting trouble spots.

E-17. **Security Illumination Mat System.** The security illumination mat system is an under vehicle illumination system. It consists of a fiber optic system that uses a single light source rated at approximately 4,000 hours. It is rapidly deployed in approximately 25 minutes. The system is capable of resisting high impact. It is also resistant to a drive over and to fuel and oil contamination.

E-18. **Security Lighted Assessment Mirror.** The security lighted assessment mirror is capable of 90-degree range of mirror movement. It consists of a continuous optical fiber mounted around the edge of the search mirror. The device has a handheld flashlight attachment and an adjustable arm support. The mirror surface is replaceable.

E-19. **Flexible Head Video Camera.** The Flexible Head Camera consists of:

- A 127mm color monitor.
- A video imaging camera with integral LED lighting built into the head of the camera.
- A four-foot cable that links the monitor to the camera.

E-20. **Expray.** The Expray is an explosive trace detector kit capable of detecting Polynitroaromatics, Nitrate Esters Nitamines, Nitrate-Based Compounds, Chlorate/Bromate-Based Explosives, and Peroxide-Based Explosives. The Expray is durable, with a shelf life of one year. There are 100 test strips in each kit. The Expray is a field test kit designed to aid in the investigation of suspects, suspicious packages, possible explosive devices, and post-blast explosive identification. Testing with the Expray is done in the following manner:

- Slide one collection paper from the dispenser.
- Peel the protective layer exposing the collection surface. Be sure not to touch the exposed collection surface to avoid contaminating the collection surface.
- Wipe the suspected area or touch suspected substance with the collection paper.
- Shake and spray Expray #1 can briefly onto the collection paper, keeping the can four inches from the paper, and observe for an immediate dark brown (similar to the color of the letter "E" on the can label) color reaction. A positive reaction indicates the presence of a group A explosive. A chart is included with the Expray kit that graphically describes the differences between the explosive classes and the interpretation of the color changes.

Components of the Checkpoint Operations Mission Module

- Next, shake and spray Expray #2 can onto the same collection paper and observe for an immediate pink (similar to the color of the letter "X" on the can label) color reaction. A positive reaction indicates the presence of a group B explosive.

- Then, shake and spray Expray #3 can onto the same collection paper and observe for an immediate pink (similar to the color of the letter "I" on the can label) color reaction. A positive reaction indicates the presence of Nitrates.

Note: Follow established guidelines when traces of explosives have been found on a suspect or suspicious package.

E-20. **Vehicle Inspection Checklists.** The vehicle inspection checklist is published by the Technical Support Working Group (TSWG) for combating terrorism. The checklist is updated periodically and is readily available through TSWG and the Government Printing Office. Other references that may be useful are the Center for Army Lessons Learned publication 07-21, The Escalation of Force Handbook, the Joint Forward Operating Base Force Protection Handbook, and TC 19-210, Access Control Handbook.

E-21. **Armorer's Kit.** The Armorer's Kit contains:

- One (1) AA Lithium Battery Four-Pack.
- Metal Detector End Caps.
- Phraselator Battery Door.
- Two (2) Phraselator Styli.
- One (1) Adjustable Wrench.
- Two (2) Phillips Screwdrivers.
- One (1) Flat Screwdriver.
- One (1) Owner's Manual, which includes the technical manuals for all the items contained in the Checkpoint Operations Mission Module.
- One (1) Pelican Case.

E-22. **Accountability.** Many of the items found in the Checkpoint Operations Mission Module will be expended or lost after use. The majority of the expendable components can be re-supplied through standard supply channels or General Services Administration purchase.

This page intentionally left blank.

Glossary

The glossary lists acronyms and terms with Army, Multi-Service, or Joint definitions, and other selected terms.

SECTION I – ACRONYMS AND ABBREVIATIONS

AI	assistant instructor
ANSEC	Army Nonlethal Scalable Effects Center
ARNG	Army National Guard
ARNGUS	Army National Guard of the United States
ASI	additional skill identifier
CALL	Center for Army Lessons Learned
CBART	capabilities based revision team
CID	central information display
CRM	Composite Risk Management
DA	Department of the Army
DoD	Department of Defense
DOTMLPF	Doctrine, Organization, Training, Materiel, Leadership and Education, Personnel, and Facilities
DPC	digital pulse controller
EMD	electro-muscular disruption
EOD	Explosive Ordnance Disposal
EOF	escalation of force
FATS	firearms training system
FM	field manual
FPS	feet per second
INIWIC	Inter-service Nonlethal Individual Weapons Instructor Course
IRCAD	individual riot control agent disperser
MAD	Magnetic Audio Device
MB	Megabyte
MCCM	modular crowd control munition
METT-TC	mission, enemy, terrain and weather, troops and support available, time available, and civil considerations
MMTSP	multi-media training support package
MOPP	Mission Oriented Protective Posture
NCO	Noncommissioned Officer
NLCS	nonlethal capabilities set
NLRT	nonlethal review team

Glossary

NLW	nonlethal weapon
OC	oleoresin capsicum
OO	Off/Off
PBIED	Personnel Borne Improvised Explosive Device
PMCS	preventive maintenance checks and services
PSI	pounds per square inch
PVAB	Portable Vehicle Arresting Barrier
ROE	rules of engagement
RSO	Range Safety Officer
RUF	rules for the use of force
SOP	standard operating procedure
TC	Training Circular
TRADOC	Training and Doctrine Command
TSWG	Technical Support Working Group
TTP	tactics, techniques, and procedures
USAR	U.S. Army Reserve
USMC	U.S. Marine Corps
VBIED	vehicle borne improvised explosive device
VLAD	Vehicle Lightweight Arresting Device
VRM	variable release mechanism
VRT	Voice Response Translator

SECTION II – TERMS

Area Target: A target consisting of an area rather than a single point. *(JP 1-02)*

Checkpoint: (DoD, NATO) A place where military police check vehicular or pedestrian traffic in order to enforce circulation control measures and other laws, orders, and regulations. See FM 3-19.4. (Army)

Convoy security operation: Specialized kind of area security operation conducted to protect convoys. (FM 3-90)

Counter-Materiel: Directed effects against materiel (vehicles, vessels, aircraft, buildings, facilities, structures, weapon systems, ammunition, and WMD, etc.). Note: Non-Lethal Counter-Materiel effects must remain non-lethal to personnel. *(Nonlethal Review Team-NLRT)*

Counter-Personnel: Directed effects against individual(s). *(NLRT)*

Device: A piece of equipment or a mechanism designed to serve a specific purpose or function. *(American Heritage)*

Disable: To render ineffective or unable to perform. *(Capabilities Based Revision Team-CBART)*

Disorient: To cause to lose one's way, perception of time, place, or one's personal identity. *(American Heritage)*

Distract: To cause to turn away from the original focus of attention or interest; divert. *(American Heritage)*

Glossary

Duration: The time an effect from a capability must last. *(CBART)*

Effect: 1. The physical or behavioral state of a system that results from an action, set of actions, or another effect. 2. The result, outcome, or consequence of an action. 3. A change to a condition, behavior or degree of freedom. *(JP 1-02)*

Environment: The circumstances, objects, or conditions by which one is surrounded. *(Merriam-Webster)*

Incapacitate: To disable, inhibit, or degrade one or more functions or capabilities of a target to render it ineffective. *(NLRT)*

Injury: A term comprising such conditions as fractures, wounds, sprains, strains, dislocations, concussions, and compressions. In addition, it includes conditions resulting from extremes of temperature or prolonged exposure. Acute poisonings (except those due to contaminated foods) resulting from exposure to a toxic or poisonous substance are also classed as injuries. *(JP 1-02)*

Large Vehicles: Semi-trailers, both boxed and bulk cargo. *(CBART)*

Lethal: Causing death or gross physical destruction. *(NLRT)*

Medium Vehicles: Small box vans up to and including water/fuel trucks. *(CBART)*

Munition: A complete device charged with explosives, propellants, pyrotechnics, initiating composition, or nuclear, biological, or chemical (NBC) material for use in military operations, including demolitions. Certain suitably modified munitions can be used for training, ceremonial, or non-operational purposes. Also called ammunition. (Note: In common usage, "munitions" [plural] can be military weapons, ammunition, and equipment). See also explosive ordnance. *(JP 1-02)*

Non-Compliant: A non-compliant target is one who exhibits passive resistance, active resistance, aggression, and/or assaultive behavior in response to a NLE or is otherwise unaffected/unaware of a NLE. *(CBART)*

Non-Lethal: Neutralizing or incapacitating a target without causing permanent injury, death, or gross physical destruction. *(NLRT)*

Non-Lethal Weapons: Weapons, devices and munitions that are explicitly designed and primarily employed to immediately incapacitate targeted personnel or materiel, while minimizing fatalities, permanent injury to personnel, and undesired damage to property in the target area or environment. Non-lethal weapons are intended to have reversible effects on personnel or materiel. *(CBART)*

Permanent Injury: Physical damage that permanently impairs physiological function that restricts employment and/or other activities of a person for the rest of his/her life. *(CBART)*

Point Target: A target of such small dimension that it requires the accurate placement of ordnance in order to neutralize or destroy it. *(FM 1-02, "Operational Terms and Graphics")*

Reversibility: The ability to return the target to its pre-engagement functionality. It is usually measured by time and level of effort required for recovery of the target. Reversibility has a direct impact on risk of permanent injury (RPI) and risk of permanent damage. *(CBART)*

Small Vehicles: Four wheeled cargo vans and smaller. *(CBART)*

Standard: Quantitative or qualitative measures for specifying the levels of performance of a task. *(CJCSM 3170.01D, February 09)*

Static: A target that is fixed in one place. *(CBART)*

Glossary

Target: An area, complex, installation, force, equipment, capability, function, or behavior identified for possible action to support the commander's objectives, guidance, and intent. *(JP 1-02)*

Task: An action or activity (derived from an analysis of the mission and concept of operations) assigned to an individual or organization to provide a capability. *(CJCSM 3170.01D, February 09)*

Vehicle: A manned or unmanned craft designed to move on land. *(CBART)*

Weapon: A means of applying force to defend against or defeat a target. *(NLRT)*

References

These are the sources used in the development of this publication.

Army Publications

AR 190-14, *Carrying of Firearms and Use of Force for Law Enforcement and Security Duties*, 12 March 1993.

AR 350-1, *Army Training and Leader Development*, 3 August 2007.

AR 350-38, *Training Device Policies and Management*, 15 October 1993.

FM 1-02, *Operational Terms and Graphics*, 21 September 2004.

FM 3-19.4, *Military Police Leaders' Handbook*, 4 March 2002.

FM 3-19.15, *Civil Disturbance Operations*, 18 April 2005.

FM 3-22.40, *Multi-Service Tactics, Techniques, and Procedures for the Tactical Employment of Nonlethal Weapons*, 24 October 2007.

FM 3-34.214, *Explosives and Demolitions*, 11 July 2007.

FM 3-90, *Tactics*, 4 July 2001.

FM 5-19, *Composite Risk Management*, 21 August 2006.

TB 9-1025-211-10, *Operator Information for the Disrupter, Non-Lethal: Taser® X26E*, 2 February 2007.

TC 19-210, *Access Control Handbook*, 4 October 2004.

TM 5-4240-342-12&P, *Operator's and Unit Maintenance Manual (Including Repair Parts and Special Tools List) for Barrier, Vehicle Arresting, Portable: M1 (NSN 4240-01-469-6122)*, 22 October 2001.

TM 5-4240-536-10, *Operator's Manual for Barrier, Vehicle Arresting, Portable: Vehicle Lightweight Arresting Device (VLAD), M2 (NSN 4240-01-518-4626)*, 30 June 2005.

Department of the Army Forms

DA Form 2028, *Recommended Changes to Publications and Blank Forms*.
DA Forms are available on the APD web site (www.apd.army.mil).

Other Publications

CALL 07-21, *Escalation of Force Handbook*, July 2007.

CJCSM 3170.01D, *Manual for the Operation of the Joint Capabilities Integration and Development System*, February 2009.

DA PAM 350-38, *Standards in Training Commission*, 13 May 2009.

DODD 3000.3, *Policy for Non-Lethal Weapons*, 9 July 1996.

Joint Capabilities Document for Joint Nonlethal Effects.

Joint Non-Lethal Weapons Directorate's Non-lethal Weapons Training and Qualification Manual.

JP 1-02, *Department of Defense Dictionary of Military and Associated Terms*, 12 April 2001.

TRADOC Pamphlet 525-99, *Concept for Nonlethal Capabilities in Army Operations*, 22 June 2005.

Multi-Media and On-line Sources

Joint Nonlethal Weapons Program, https://www.jnlwp.com/.

This page intentionally left blank.

Index

Accuracy 2-3, 8-3, 8-6, 9-13, 11-3, 11-8
Aiming point 2-2, 3-3, 3-5,4-4, 4-5, 6-5, 6-6
Anti-felon identification tags 9-8
Armorer's Kit A-3

Caltrops 10-1, 10-13
Central Information Display 9-2, 9-4
Counter-Materiel 1-2, 10-10
Counter-Personnel 1-2

Digital Pulse Controller 9-4

Electro-muscular disruption 9-1
Engagement skills trainer 2-3
Escalation of force 1-2, 1-4, 2-7
Expray E-4, 5

Force continuum 2-6, 2-7

Hangfire 7-3
Hydraulic needle effect 8-4

Igniter 6-1, 6-2, 6-3, 6-4, 6-6, 6-7
Individual Riot Control Agent Disperser .. 8-1, 8-3

Joint Nonlethal Weapons Program 1-3

M1006 Certification 4-2
M1006 Sighting 4-2, 4-3
M1012 Certification 3-2
M1012 Sighting 3-3
M1013 Familiarization 3-4

M1013 Sighting 3-5
M1029 Familiarization 4-5
M1029 Sighting 4-5
Magnetic Audio Device 11-1, 11-11, 11-12
Magnum spike 10-1, 10-13
Misfire 6-5, 6-6, 6-7,7-3, 9-13
Modular Accessory Shotgun System 3-1
Modular Crowd Control Munition 2-3, 6-1
Mossberg 500 3-1
Multi-Media Training Support Package 1-1, 1-3

Nonlethal Advisor 1-3, 2-1
Nonlethal Capabilities Set 1-2, A-1
Nonlethal grenades 2-3, 5-1, 7-1

Pepper Spray 8-2
Portable Vehicle Arresting Barrier 10-1

Rules for the Use of Force 2-6
Rules of engagement 1-1, 1-2, 2-6

Shock tube 6-1, 6-2, 6-3, 6-4, 6-5, 6-6, 6-7
Sponge Grenade 2-3, 2-4, 4-1, 4-2
Spray patterns 8-3, 8-7
Sting Ball Grenade 5-1, 5-4, 5-5
Stun Hand Grenade 5-1, 5-2
Sweet spot 2-2, 2-3, 3-1, 3-3, 4-1, 4-4, 6-4

Vehicle arresting device 10-1, 10-13
Vehicle Barrier 10-1
Vehicle Lightweight Arresting Device ... 1-2,10-1, 10-10
Voice Response Translator 11-1, 11-9

This page intentionally left blank.

TC 3-19.5
5 November 2009

By Order of the Secretary of the Army:

GEORGE W. CASEY, JR.
General, United States Army
Chief of Staff

Official:

JOYCE E. MORROW
Administrative Assistant to the
Secretary of the Army
0928802

DISTRIBUTION:
Active Army, Army National Guard, and United States Army Reserve: Not to be distributed; electronic media only.

PIN: 085848-000

www.ingramcontent.com/pod-product-compliance
Lightning Source LLC
Chambersburg PA
CBHW050104230526
45470CB00004B/1666